Mike Holt's Illustrated Guide to

Understanding the NEC® Requirements for
Grounding
versus Bonding

2008 Edition

MIKE HOLT ENTERPRISES, INC.

TM

Since 1974
www.MikeHolt.com

Mike Holt Enterprises, Inc.
1.888.NEC.CODE • www.MikeHolt.com • Info@MikeHolt.com

NOTICE TO THE READER

Mike Holt's Illustrated Guide to
Understanding the NEC Requirements for Grounding versus Bonding

First Printing: January 2008

Technical Illustrator: Mike Culbreath
Cover Design: Tracy Jette
Layout Design and Typesetting: Cathleen Kwas and Tara Moffitt

COPYRIGHT © 2008 Charles Michael Holt, Sr.
ISBN 978-1-932685-38-1

For more information, call 1.888.NEC.CODE (888-632-2633), or E-mail Info@MikeHolt.com.

This logo is a registered trademark of Mike Holt Enterprises, Inc.

To request examination copies of this or other Mike Holt Publications, call:
Phone: 1.888.NEC.CODE (888-632-2633) • Fax: 1.352.360.0983
or E-mail: Info@MikeHolt.com
or visit Mike Holt online: www.MikeHolt.com
You can download a sample PDF of all our publications by visiting www.MikeHolt.com

I dedicate this book to the
Lord Jesus Christ,
my mentor and teacher.
Proverbs 16:3

One Team

To Our Instructors and Students:

We are committed to providing you the finest product with the fewest errors, but we are realistic and know that there will be errors found and reported after the printing of this book. The last thing we want is for you to have problems finding, communicating, or accessing this information. It is unacceptable to us for there to be even one error in our textbooks or answer keys. For this reason, we are asking you to work together with us as One Team.

Students: Please report any errors that you may find to your instructor.
Instructors: Please communicate these errors to us.

Our Commitment:

We will continue to list all of the corrections that come through for all of our textbooks and answer keys on our Website. We will always have the most up-to-date answer keys available to instructors to download from our instructor Website. We do not want you to have problems finding this updated information, so we're outlining where to go for all of this below:

To view textbook and answer key corrections: Students and instructors go to our Website, www.MikeHolt.com, click on "Books" in the sidebar of links, and then click on "Corrections."

To download the most up-to-date answer keys: Instructors go to our Website, www.MikeHolt.com, click on "Instructors" in the sidebar of links and then click on "Answer Keys." On this page you will find instructions for accessing and downloading these answer keys.

If you are not registered as an instructor you will need to register. Your registration will be sent to our educational director who in turn reviews and approves your registration. In your approval E-mail will be the login and password so you can have access to all of the answer keys. If you have a situation that needs immediate attention, please contact the office directly at 1.888.NEC.CODE.

Call 1.888.NEC.CODE or visit us online at www.MikeHolt.com

Table of Contents

Introduction

Mike Holt's Illustrated Guide to Understanding the NEC Requirements for Grounding versus Bonding

This textbook covers the Grounding and Bonding requirements contained in the *National Electrical Code*. Grounding is the most important and least understood article in the *NEC*, and surveys have repeatedly shown that the majority of electrical shocks power quality problems are due to improper grounding or bonding. The writing style of this textbook, and in all of Mike Holt's products, is meant to be informative, practical, useful, informal, easy to read, and applicable for today's electrical professional. Also, just like all of Mike Holt's textbooks, it contains hundreds of full-color illustrations to help you see the safety requirements of the *National Electrical Code* in practical use, as they apply to today's electrical installations.

This illustrated textbook contains advice, cautions about possible conflicts or confusing *Code* requirements, tips on proper electrical installations, and warnings of dangers related to improper electrical installations. In spite of this effort, some rules are unclear or need additional editorial improvement.

This textbook can't eliminate confusing, conflicting, or controversial *Code* requirements, but it does try to put these requirements into sharper focus to help you understand their intended purpose. Sometimes a requirement is so confusing nobody really understands its actual application. When this occurs, this textbook will point the situation out in an up front and straightforward manner.

The *NEC* is updated every three years to accommodate new electrical products and materials, changing technologies, and improved installation techniques, along with editorial improvements. While the uniform adoption of each new edition of the *Code* is the best approach for all involved in the electrical industry, many inspection jurisdictions modify the *NEC* when it's adopted. In addition, the *Code* allows the authority having jurisdiction, also known as the "AHJ," typically the electrical inspector, the authority to waive *NEC* requirements or permit alternative wiring methods contrary to the *Code* requirements when assured the completed electrical installation is equivalent in establishing and maintaining effective safety [90.4].

Keeping up with the *NEC* should be the goal of all those who are involved in the safety of electrical installations. This includes electrical installers, contractors, owners, inspectors, engineers, instructors, and others concerned with electrical installations.

About the 2008 *NEC*

The actual process of changing the *Code* takes about two years, and it involves thousands of individuals making an effort to have the *NEC* as current and accurate as possible. Let's review how this process works:

Step 1. *Proposals—November, 2005.* Anybody can submit a proposal to change the *Code* before the proposal closing date. Over 3,600 proposals were submitted to modify the 2008 *NEC*; of these proposals, 300 rules were revised that significantly affect the electrical industry. Some changes were editorial revisions, while others were more significant, such as new Articles, Sections, Exceptions, and Fine Print Notes.

Step 2. *Code Making Panels Review Proposals—January, 2006.* All *Code* proposals were reviewed by Code Making Panels (there were 20 panels in the 2008 *Code* process) who voted to accept, reject, or modify them.

Step 3. *Report on Proposals (ROP)—July, 2006.* The voting of the 20 Code Making Panels on the proposals was published for public review in a document called the "Report on Proposals," frequently referred to as the "ROP."

Step 4. *Public Comments—October, 2006.* Once the ROP was available, public comments were submitted asking the Code Making Panels members to revise their earlier actions on change proposals, based on new information. The closing date for "Comments" was October, 2006.

Step 5. *Comments Reviewed by Code Making Panels—December, 2006.* The Code Making Panels met again to review, discuss, and vote on public comments.

Step 6. *Report on Comments (ROC)—April, 2007.* The voting on the "Comments" was published for public review in a document called the "Report on Comments," frequently referred to as the "ROC."

Step 7. *Electrical Section—June, 2007.* The NFPA Electrical Section discussed and reviewed the work of the Code Making Panels. The Electrical Section developed recommendations on last-minute motions to revise the proposed *NEC* draft that would be presented at the NFPA annual meeting.

Step 8. *NFPA Annual Meeting—June, 2007.* The 2008 *NEC* was officially adopted at the annual meeting, after a number of motions (often called "floor actions") were voted on.

Step 9. *Standards Council Review Appeals and Approves the 2008 NEC—July, 2007.* The NFPA Standards Council reviewed the record of the *Code*-making process and approved publication of the 2008 *NEC*.

Step 10. *2008 NEC Published—September, 2007.* The 2008 *National Electrical Code* was published, following the NFPA Board of Directors review of appeals.

> **Author's Comment:** Submitting proposals and comments online can be accomplished by going to the NFPA Website (www.nfpa.org), click on "Codes and Standards" at the top of the page, and once on the Codes and Standards page click on "Proposals and Comments" in the box on the right hand side of the page. The deadline for proposals to create the 2011 *National Electrical Code* is November 5, 2008.

The Scope of this Textbook

This textbook, Mike Holt's *Illustrated Guide to Understanding the NEC Requirements for Grounding versus Bonding,* covers the general installation requirements contained in the *NEC* from Article 90 through 820 that Mike considers to be of critical importance.

But the textbook contains the following stipulations:

- **Power Systems and Voltage.** All power supply systems are assumed to be solidly grounded and of any of the following voltages: 120V single-phase, 120/240V single-phase, 120/208V three-phase, 120/240V three-phase, or 277/480V three-phase, unless identified otherwise.

- **Electrical Calculations.** Unless the question or example specifies three-phase, the questions and examples are based on a single-phase power supply.

- **Rounding.** All calculations are rounded to the nearest ampere in accordance with 220.5(B).

- **Conductor Material.** All conductors are considered copper, unless aluminum is identified or specified.

- **Conductor Sizing.** All conductors are sized based on a THHN copper conductor terminating on a 75°C terminal in accordance with 110.14(C), unless the question or example identifies otherwise.

- **Overcurrent Device.** The term "overcurrent device" in this textbook refers to a molded case circuit breaker, unless identified otherwise. Where a fuse is identified, it's to be of the single-element type, also known as a "onetime fuse," unless identified otherwise.

How to Use This Textbook

This textbook is to be used with the *NEC*, not as a replacement for the *Code* book, so be sure to have a copy of the 2008 *National Electrical Code* handy. Compare what Mike is explaining to the text in your *Code* book, and discuss those topics that you find difficult to understand with others.

You'll notice that all *NEC* text has been paraphrased, as well as some of the Articles and Sections titles are different than they appear in the actual *Code*. This only occurs when Mike believes it's easier to understand the content of the rule, so keep this in mind when comparing this textbook against the actual *NEC*.

As you read through this textbook, be sure to take the time to review the text along with the outstanding graphics and examples provided.

Textbook Format

This textbook follows the *NEC* format, but it doesn't cover every *Code* requirement. For example, it doesn't include every Article, Section, Subsection, Exception, or Fine Print Note. So don't be concerned if you see the textbook contain Exception No. 1 and Exception No. 3, but not Exception No. 2.

Graphics with red borders are graphics that contain a 2008 change; graphics without a red border are graphics that support the concept being discussed, but nothing in the graphic was affected by a 2008 *Code* change.

Special Sections and Examples. Additional information to better help you understand a concept is identified with light green shading. In addition, examples are highlighted with a yellow background.

Cross-References and Author's Comments

This textbook contains several *NEC* cross-references to other related *Code* requirements to help you develop a better understanding of how the *NEC* rules relate to one another. These cross-references are identified by *Code* Section numbers in brackets, an example of which is "[90.4]."

Author's comments were written by Mike to help you the reader better understand the *NEC* by bringing to your attention items you should be aware of.

Difficult Concepts

As you progress through this textbook, you might find you don't understand every explanation, example, calculation, or comment. Don't get frustrated, and don't get down on yourself. Remember, this is the *National Electrical Code,* and sometimes the best attempt to explain a concept isn't enough to make it perfectly clear. If you're still confused, visit www. MikeHolt.com, and post your question on the *Code* Forum for help.

Different Interpretations

Some electricians, contractors, instructors, inspectors, engineers, and others enjoy the challenge of discussing the *Code* requirements, hopefully in a positive and a productive manner. This give-and-take is important to the process of better understanding *NEC* requirements and applications. However, if you're going to get into an *NEC* discussion, please don't spout out what you think without having the actual *Code* in your hand. The professional way of discussing an *NEC* requirement is by referring to a specific Section, rather than talking in vague generalities.

Textbook Errors and Corrections

If you believe there's an error of any kind in this textbook (typographical, grammatical, or technical), no matter how insignificant, please let us know.

Any errors found after printing are listed on our Website, so if you find an error, first check to see if it has already been corrected. Go to www. MikeHolt.com, click on the "Books" link, and then the "Corrections" link (www. MikeHolt.com/bookcorrections.htm).

If you don't find the error listed on the Website, contact us by E-mailing us at Corrections@MikeHolt.com. Be sure to include the book title, page number, and any other pertinent information.

Grounding versus Bonding Library

To understand ground- versus bonding in greater detail, watch Mike explain this difficult to understand article. Order the 2008 Grounding versus Bonding DVDs by calling 1.888.NEC.CODE, or visiting www.MikeHolt.com.

How to Use the *National Electrical Code*

The *National Electrical Code* is written for persons who understand electrical terms, theory, safety procedures, and electrical trade practices. These individuals include electricians, electrical contractors, electrical inspectors, electrical engineers, designers, and other qualified persons. The *Code* is not written to serve as an instructive or teaching manual for untrained individuals [90.1(C)].

Learning to use the *NEC* is somewhat like learning to play the game of chess; it's a great game if you enjoy mental warfare. When learning to play chess, you must first learn the names of the game pieces, how the pieces are placed on the board, and how each piece moves.

Once you understand the fundamentals of the game of chess, you're ready to start playing the game. Unfortunately, at this point all you can do is make crude moves, because you really don't understand how all the information works together. To play chess well, you'll need to learn how to use your knowledge by working on subtle strategies before you can work your way up to the more intriguing and complicated moves.

Not a Game

Electrical work isn't a game, and it must be taken very seriously. Learning the basics of electricity, important terms and concepts, as well as the basic layout of the *NEC* gives you just enough knowledge to be dangerous. There are thousands of specific and unique applications of electrical installations, and the *Code* doesn't cover every one of them. To safely apply the *NEC*, you must understand the purpose of a rule and how it affects the safety aspects of the installation.

NEC Terms and Concepts

The *NEC* contains many technical terms, so it's crucial for *Code* users to understand their meanings and their applications. If you don't understand a term used in a *Code* rule, it will be impossible to properly apply the *NEC* requirement. Be sure you understand that Article 100 defines the terms that apply to two or more Articles. For example, the term "Dwelling Unit" applies to many Articles. If you don't know what a dwelling unit is, how can you apply the *Code* requirements for it?

In addition, many Articles have terms unique for that specific Article. This means that the definitions of those terms are only applicable for that given Article. For example, Section 250.2 contains the definitions of terms that only apply to Article 250, Grounding and Bonding.

Small Words, Grammar, and Punctuation

It's not only the technical words that require close attention, because even the simplest of words can make a big difference to the intent of a rule. The word "or" can imply alternate choices for equipment wiring methods, while "and" can mean an additional requirement. Let's not forget about grammar and punctuation. The location of a comma "," can dramatically change the requirement of a rule.

Slang Terms or Technical Jargon

Electricians, engineers, and other trade-related professionals use slang terms or technical jargon that isn't shared by all. This makes it very difficult to communicate because not everybody understands the intent or application of those slang terms. So where possible, be sure you use the proper word, and don't use a word if you don't understand its definition and application. For example, lots of electricians use the term "pigtail" when describing the short conductor for the connection of a receptacle, switch, luminaire, or equipment. Although they may understand this, not everyone does.

NEC Style and Layout

Before we get into the details of the *NEC*, we need to take a few moments to understand its style and layout. Understanding the structure and writing style of the *Code* is very important before it can be used effectively. If you think about it, how can you use something if you don't know how it works? The *National Electrical Code* is organized into ten components.

1. Table of Contents
2. Article 90 (Introduction to the *Code*)
3. Chapters 1 through 9 (major categories)
4. Articles 90 through 830 (individual subjects)
5. Parts (divisions of an Article)

6. Sections and Tables (*Code* requirements)
7. Exceptions (*Code* permissions)
8. Fine Print Notes (explanatory material)
9. Annexes (information)
10. Index

1. Table of Contents. The Table of Contents displays the layout of the Chapters, Articles, and Parts as well as the page numbers. It's an excellent resource and should be referred to periodically to observe the interrelationship of the various *NEC* components. When attempting to locate the rules for a particular situation, knowledgeable *Code* users often go first to the Table of Contents to quickly find the specific *NEC* Part that applies.

2. Introduction. The *NEC* begins with Article 90, the introduction to the *Code*. It contains the purpose of the *NEC*, what is covered and what is not covered along with how the *Code* is arranged. It also gives information on enforcement and how mandatory and permissive rules are written as well as how explanatory material is included. Article 90 also includes information on formal interpretations, examination of equipment for safety, wiring planning, and information about formatting units of measurement.

3. Chapters. There are nine Chapters, each of which is divided into Articles. The Articles fall into one of four groupings: General Requirements (Chapters 1 through 4), Specific Requirements (Chapters 5 through 7), Communications Systems (Chapter 8), and Tables (Chapter 9).

- Chapter 1 General
- Chapter 2 Wiring and Protection
- Chapter 3 Wiring Methods and Materials
- Chapter 4 Equipment for General Use
- Chapter 5 Special Occupancies
- Chapter 6 Special Equipment
- Chapter 7 Special Conditions
- Chapter 8 Communications Systems (Telephone, Data, Satellite, and Cable TV)
- Chapter 9 Tables–Conductor and Raceway Specifications

4. Articles. The *NEC* contains approximately 140 Articles, each of which covers a specific subject. For example:

- Article 110 General Requirements
- Article 250 Grounding and Bonding
- Article 300 Wiring Methods
- Article 430 Motors and Motor Controllers
- Article 500 Hazardous (Classified) Locations
- Article 680 Swimming Pools, Fountains, and Similar Installations

- Article 725 Remote-Control, Signaling, and Power-Limited Circuits
- Article 800 Communications Systems

5. Parts. Larger Articles are subdivided into Parts.

Author's Comment: Because the Parts of a *Code* article aren't included in the section numbers, we have a tendency to forget what "Part" the *NEC* rule is relating to. For example, Table 110.34(A) contains the working space clearances for electrical equipment. If we aren't careful, we might think this table applies to all electrical installations, but Table 110.34(A) is located in Part III, which contains the requirements for Over 600 Volts, Nominal installations. The rules for working clearances for electrical equipment for systems 600V, Nominal, or less are contained in Table 110.26(A)(1), which is located in Part II—600 Volts, Nominal, or Less.

6. Sections and Tables.

Sections. Each *NEC* rule is called a *Code* Section. A *Code* Section may be broken down into subsections by letters in parentheses "(A), (B)," etc. Numbers in parentheses (1), (2), etc., may further break down a subsection, and lowercase letters (a), (b), etc., further break the rule down to the third level. For example, the rule requiring all receptacles in a dwelling unit bathroom to be GFCI-protected is contained in Section 210.8(A)(1). Section 210.8(A)(1) is located in Chapter 2, Article 210, Section 8, subsection (A), sub-subsection (1).

Many in the industry incorrectly use the term "Article" when referring to a *Code* Section. For example, they say "Article 210.8," when they should say "Section 210.8."

Tables. Many *Code* requirements are contained within Tables, which are lists of *NEC* requirements placed in a systematic arrangement. The titles of the Tables are extremely important; you must read them carefully in order to understand the contents, applications, limitations, etc., of each Table in the *Code*. Many times notes are provided in or below a Table; be sure to read them as well since they are also part of the requirement. For example, Note 1 for Table 300.5 explains how to measure the cover when burying cables and raceways, and Note 5 explains what to do if solid rock is encountered.

7. Exceptions. Exceptions are *Code* requirements or allowances that provide an alternative method to a specific requirement. There are two types of exceptions—mandatory and permissive. When a rule has several exceptions, those exceptions with mandatory requirements are listed before the permissive exceptions.

Mandatory Exception. A mandatory exception uses the words "shall" or "shall not." The word "shall" in an exception means that if you're using the exception, you're required to do it in a particular way. The phrase "shall not" means it isn't permitted.

Permissive Exception. A permissive exception uses words such as "shall be permitted," which means it's acceptable (but not mandatory) to do it in this way.

8. Fine Print Note (FPN). A Fine Print Note contains explanatory material intended to clarify a rule or give assistance, but it isn't a *Code* requirement [90.5(C)].

9. Annexes. Annexes aren't a part of the *NEC* requirements, and are included in the *Code* for informational purposes only.

- Annex A. Product Safety Standards
- Annex B. Application Information for Ampacity Calculation
- Annex C. Raceway Fill Tables for Conductors and Fixture Wires of the Same Size
- Annex D. Examples
- Annex E. Types of Construction
- Annex F. Critical Operations Power Systems (COPS)
- Annex G. Supervisory Control and Data Acquisition (SCADA)
- Annex H. Administration and Enforcement

10. Index. The Index at the back of the *NEC* is helpful in locating a specific rule.

Author's Comment: Changes to the *NEC* since the previous edition(s), are identified by shading, but rules that have been relocated aren't identified as a change. A bullet symbol "•" is located on the margin to indicate the location of a rule that was deleted from a previous edition.

How to Locate a Specific Requirement

How to go about finding what you're looking for in the *Code* depends, to some degree, on your experience with the *NEC*. *Code* experts typically know the requirements so well they just go to the correct rule without any outside assistance. The Table of Contents might be the only thing very experienced *NEC* users need to locate the requirement they're looking for. On the other hand, average *Code* users should use all of the tools at their disposal, and that includes the Table of Contents and the Index.

Table of Contents. Let's work out a simple example: What *NEC* rule specifies the maximum number of disconnects permitted for a service? If you're an experienced *Code* user, you'll know Article 230 applies to "Services," and because this article is so large, it's divided up into multiple Parts (actually eight Parts). With this knowledge, you can quickly go to the Table of Contents and see that it lists Service Equipment Disconnecting Means requirements in Part VI.

Author's Comment: The number 70 precedes all page numbers because the *NEC* is NFPA standard number 70.

Index. If you use the Index, which lists subjects in alphabetical order, to look up the term "service disconnect," you'll see there's no listing. If you try "disconnecting means," then "services," you'll find the Index specifies the rule is located at Article 230, Part VI. Because the *NEC* doesn't give a page number in the Index, you'll need to use the Table of Contents to find the page number, or flip through the *Code* to Article 230, then continue to flip through pages until you find Part VI.

Many people complain that the *NEC* only confuses them by taking them in circles. As you gain experience in using the *Code* and deepen your understanding of words, terms, principles, and practices, you will find the *NEC* much easier to understand and use than you originally thought.

Customizing Your *Code* Book

One way to increase your comfort level with the *Code* is to customize it to meet your needs. You can do this by highlighting and underlining important *NEC* requirements, and by attaching tabs to important pages.

Highlighting. As you read through this textbook, be sure you highlight those requirements in the *Code* that are the most important or relevant to you. Use yellow for general interest and orange for important requirements you want to find quickly. Be sure to highlight terms in the Index and Table of Contents as you use them.

Underlining. Underline or circle key words and phrases in the *NEC* with a red pen (not a lead pencil) and use a six-inch ruler to keep lines straight and neat. This is a very handy way to make important requirements stand out. A small six-inch ruler also comes in handy for locating specific information in the many *Code* tables.

Tabbing the *NEC.* By placing tabs on *Code* Articles, Sections, and Tables, it will make it easier for you to use the *NEC*. However, too many tabs will defeat the purpose. You can order a custom set of *Code* tabs online at www.MikeHolt.com, or by calling 1.888.NEC.CODE.

About the Author

Mike Holt worked his way up through the electrical trade from an apprentice electrician to become one of the most recognized experts in the world as it relates to electrical power installations. He was a Journeyman Electrician, Master Electrician, and Electrical Contractor. Mike came from the real world, and he has a unique understanding of how the *NEC* relates to electrical installations from a practical standpoint. You'll find his writing style to be simple, nontechnical, and practical.

Did you know that he didn't finish high school? So if you struggled in high school or if you didn't finish it at all, don't let this get you down, you're in good company. As a matter of fact, Mike Culbreath, Master Electrician, who produces the finest electrical graphics in the history of the electrical industry, didn't finish high school either. So two high school dropouts produced the text and graphics in this textbook! However, realizing success depends on one's continuing pursuit of education, Mike immediately attained his GED (as did Mike Culbreath) and ultimately attended the University of Miami's Graduate School for a Master's degree in Business Administration (MBA).

Mike Holt resides in Central Florida, is the father of seven children, and has many outside interests and activities. He is a five-time National Barefoot Water-Ski Champion (1988, 1999, 2005, 2006, and 2007); he has set many national records and continues to train year-round at a World competition level [www.barefootwaterskier.com].

What sets him apart from some is his commitment to living a balanced lifestyle; he places God first, then family, career, and self.

Acknowledgments

Special Acknowledgments

First, I want to thank God for my godly wife who is always by my side and my children, Belynda, Melissa, Autumn, Steven, Michael, Meghan, and Brittney.

A special thank you must be sent to the staff at the National Fire Protection Association (NFPA), publishers of the *NEC*—in particular Jeff Sargent for his assistance in answering my many *Code* questions over the years. Jeff, you're a "first class" guy, and I admire your dedication and commitment to helping others understand the *NEC*. Other former NFPA staff members I would like to thank include John Caloggero, Joe Ross, and Dick Murray for their help in the past.

A personal thank you goes to Sarina, my long-time friend and office manager. It has been wonderful working side-by-side with you for over 25 years nurturing this company's growth from its small beginnings.

Mike Holt Enterprises Team

Graphic Illustrator

Mike Culbreath devoted his career to the electrical industry and worked his way up from an apprentice electrician to master electrician. While working as a journeyman electrician, he suffered a serious on-the-job knee injury. With a keen interest in continuing education for electricians, he completed courses at Mike Holt Enterprises, Inc. and then passed the exam to receive his Master Electrician's license. In 1986, after attending classes at Mike Holt Enterprises, Inc. he joined the staff to update material and later studied computer graphics and began illustrating Mike Holt's textbooks and magazine articles. He's worked with the company for over 20 years and, as Mike Holt has proudly acknowledged, has helped to transform his words and visions into lifelike graphics.

Technical Editorial Director

Steve Arne has been involved in the electrical industry since 1974 working in various positions from electrician to full-time instructor and department chair in technical postsecondary education. Steve has developed curriculum for many electrical training courses and has developed university business and leadership courses. Currently, Steve offers occasional exam prep and continuing education *Code* classes. Steve believes that as a teacher he understands the joy of helping others as they learn and experience new insights. His goal is to help others understand more of the technological marvels that surround us. Steve thanks God for the wonders of His creation and for the opportunity to share it with others.

Steve and his lovely wife Deb live in Rapid City, South Dakota where they are both active in their church and community. They have two grown children and five grandchildren.

Technical *Code* Consultant

Ryan Jackson is a combination inspector for Draper City, Utah. He is certified as a building, electrical, mechanical, and plumbing inspector. He's also certified as a building plans examiner and electrical plans examiner. Ryan is the senior electrical inspector for Draper City, and also teaches seminars on the *NEC*. Ryan is very active in the Utah Chapter of IAEI, where he is currently president. He also enjoys staying active in the *NEC* change process, and loves to help people with their *Code* problems. On Mike Holt's *Code* Forum, he has been involved in nearly 5,000 topics.

Ryan enjoys reading, going to college football games, and spending time with his wife Sharie and their two children, Kaitlynn and Aaron.

Editorial Team

I would like to thank Toni Culbreath and Barbara Parks who worked tirelessly to proofread and edit the final stages of this publication. Their attention to detail and dedication to this project is greatly appreciated.

Production Team

I would like to thank Tara Moffitt and Cathleen Kwas who worked as a team to do the layout and production of this book. Their desire to create the best possible product for our customers is appreciated.

Advisory Committee

Rahe Loftin P.E.
Fire Protection Engineer
U.S. General Services Administration
Fort Worth, TX

Video Team Members

Daniel Craft
Electrical Foreman
Westminster, CA

Daniel Craft a high school dropout had little confidence in amounting to anything of value in society. He found himself with little discipline and even less education. Daniel soon found himself desperate for something, anything. As a hard worker he had no problem keeping jobs that required a strong back.

At the young age of 20, he was offered a job as an electrician's helper, where he learned fast that if he was going to survive in this technical world, he would have to go back to school. But where would he start? Realizing the necessary sacrifice of laziness and ignorance, for a life of dedication and discipline, he began taking electrical courses at Los Angeles Trade Tech.

While working as an electrician's helper he applied for an apprenticeship with the International Brotherhood of Electrical Workers, Local 441 in Orange County, California. Over a year had passed from applying, before being accepted in the five year apprenticeship where he would complete structured electrical courses as well as extensive hands on training.

Daniel is now recognized by the IBEW and the state of California as being a Journeyman Electrician. He currently works as an Electrical Foreman for R.L. Douglass Electric, Inc. in Orange County, CA., and starting January 2008 will begin teaching continuing education courses at the Orange County Electrical Training Trust in Santa Ana, CA.

At the age of twenty-eight he has held fast, excelling in areas such as electrical construction and the *National Electrical Code*. He believes it's crucial that one must maintain the position of a student if he/she is going to be successful, especially in the electrical industry.

Daniel has a passion for helping others achieve confidence to grow into the person that God meant them to be. Contrary to the "dropout" Daniel now lives his life putting God first, family second, and career a close third.

Scott Harding
Electrical Contractor,
F.B. Harding, Inc.
Rockville, MD
http://fbharding.com

Scott Harding began his electrical career as an electrician's helper working part time in the family business. After graduating with an Electrical Engineering degree from Clemson University, he worked as an electrical engineer for RTKL Associates, a large design firm located in Baltimore, MD.

After leaving RTKL, Scott went to work at F.B. Harding, Inc. an Electrical Contractor located in Rockville, MD where he's been for the last 19 years. He is a licensed Master Electrician in multiple jurisdictions and is the current President and CEO of F.B. Harding, Inc, making the family business a third generation company.

Scott also serves on *NEC* Code-Making Panel Number 5 (Grounding and Bonding) and worked on the 2005 and 2008 cycles.

Eric Stromberg
Electrical Engineer/Instructor
Dow Chemical
Lake Jackson, TX

Eric Stromberg enrolled in the University of Houston in 1976, with Electrical Engineering as his major. During the first part of his college years, Eric worked for a company that specialized in installing professional sound systems. Later, he worked for a small electrical company and eventually became a journeyman electrician.

After graduation from college in 1982, Eric went to work for an electronics company that specialized in fire alarm systems for high-rise buildings. He became a state licensed fire alarm installation superintendent and was also a member of IBEW local union 716. In 1989, Eric began a career with The Dow Chemical Company as an Electrical Engineer designing power distribution systems for large industrial facilities. In 1997 Eric began teaching *National Electrical Code* classes.

Eric currently resides in Lake Jackson, Texas, with his wife Jane and three children: Ainsley, Austin, and Brieanna.

John Travers
Electrical Inspector and Instructor,
Electric Code Class
Hialeah, FL
http://electriccodeclass.com

John Travers completed the Joint NECA/IBEW Apprenticeship-Training program in 1977, after a tour of duty in the Air Force. He worked at various jobsites in the South Florida and Illinois areas until becoming a contractor in 1984. He began his career as an inspector in 1985 serving Hialeah Gardens. In 1988, he added the Town of Medley to his responsibilities, and became a full-time Chief Electrical Inspector for the City of Hialeah in 1990. He served in that position until 2005. He now serves as the Director of Community Development for Hialeah. His responsibilities include managing the Building, Zoning, Code Compliance and Occupational License Divisions of the City of Hialeah.

He joined the IAEI in June of 1988 and was the project chairman for the 1996 Annual Meeting in Key West, Florida. He currently serves as the Treasurer and Director of Codes & Education for the Miami-Dade Division. John oversees an aggressive Continuing Education program for the electrical industry in the South Florida area, where he is a certified Education Sponsor for electrical contractors, inspectors, engineers and tradesmen. In 2003, he began teaching nationwide, assisting the I.C.C. with seminars in Denver and Salt Lake City.

John is a family man, who has been married to Barbara for over 35 years, with four children and a half dozen grandchildren. Both his sons are currently serving our country in the Air Force. His hobbies include fishing, diving, philately, golf and photography. He is a martial artist in GOJU Karate, where he received his third degree black belt in 2006.

Laura Vergeront
IBEW Journeyman/State Licensed
Electrician/Vocational Instructor
Riverside, CA
Vergeront@msn.com

After Laura completed the Joint NECA/IBEW Apprenticeship-Training program in 1986, she worked as a journeyman electrician in new construction for 8 years. She was then hired by the California Department of Corrections to head the electrical maintenance department of Chino Men's Prison, a position she has held for 13 years.

For the last two years she has been instructing inmates full-time at the California Rehabilitation Center, Norco California in the basic safety and electrical skills needed to acquire a job once they are released.

At night her passion for the last ten years has been in teaching "Bonding and Grounding" to apprentices at the NECA/IBEW Apprenticeship School in her area.

Saturdays are filled with OSHA-10 certification classes that she teaches to apprentices and journeyman.

Laura has been married to her handsome husband Mike for 29 years. She and Mike share hobbies that include fly-fishing, backpacking and photography.

Steve Arne and **Mike Culbreath** (members of the Mike Holt Enterprises Team) were also video team members.

Notes

90

Introduction to the *National Electrical Code*

INTRODUCTION TO ARTICLE 90—INTRODUCTION TO THE *NATIONAL ELECTRICAL CODE*

Many *NEC* violations and misunderstandings wouldn't occur if people doing the work simply understood Article 90. For example, many people see *Code* requirements as performance standards. In fact, the *NEC* requirements are bare minimums for safety. This is exactly the stance electrical inspectors, insurance companies, and courts take when making a decision regarding electrical design or installation.

Article 90 opens by saying the *NEC* isn't intended as a design specification or instruction manual. The *National Electrical Code* has one purpose only. That is "the practical safeguarding of persons and property from hazards arising from the use of electricity." It goes on to indicate that the *Code* isn't intended as a design specification or instruction manual. Yet, the necessity to study, study, and study the *NEC* rules some more can't be overemphasized. Understanding where to find the rules in the *Code* that apply to the installation is invaluable. Rules in several different articles often apply to even a simple installation.

Article 90 then describes the scope and arrangement of the *NEC*. A person who says, "I can't find anything in the *Code*," is really saying, "I never took the time to review Article 90." The balance of Article 90 provides the reader with information essential to understanding those items you do find in the *NEC*.

Typically, electrical work requires you to understand the first four chapters of the *Code* which apply generally, plus have a working knowledge of the Chapter 9 tables. Chapters 5, 6, and 7 make up a large portion of the *NEC*, but they apply to special occupancies, special equipment, or other special conditions. They build on, modify, or amend the rules in the first four chapters. That knowledge begins with Article 90.

90.1 Purpose of the *NEC*.

(A) Practical Safeguarding. The purpose of the *NEC* is to ensure electrical systems are installed in a manner that protects people and property by minimizing the risks associated with the use of electricity.

(B) Adequacy. The *Code* contains requirements considered necessary for a safe electrical installation. When an electrical installation is installed in compliance with the *NEC*, it will be essentially free from electrical hazards. The *Code* is a safety standard, not a design guide.

NEC requirements aren't intended to ensure the electrical installation will be efficient, convenient, adequate for good service, or suitable for future expansion. Specific items of concern, such as electrical energy management, maintenance, and power quality issues aren't within the scope of the *Code*.
Figure 90–1

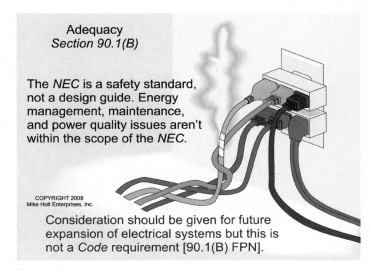

Adequacy
Section 90.1(B)

The *NEC* is a safety standard, not a design guide. Energy management, maintenance, and power quality issues aren't within the scope of the *NEC*.

COPYRIGHT 2008
Mike Holt Enterprises, Inc.

Consideration should be given for future expansion of electrical systems but this is not a *Code* requirement [90.1(B) FPN].

Figure 90–1

FPN: Hazards in electrical systems often occur because circuits are overloaded or not properly installed in accordance with the *NEC*. The initial wiring often did not provide reasonable provisions for system changes or for the increase in the use of electricity.

Author's Comments:

- See the definition of "Overload" in Article 100.

- The *NEC* does not require electrical systems to be designed or installed to accommodate future loads. However, the electrical designer, typically an electrical engineer, is concerned with not only ensuring electrical safety (*Code* compliance), but also ensuring the system meets the customers' needs, both of today and in the near future. To satisfy customers' needs, electrical systems must be designed and installed above the minimum requirements contained in the *NEC*.

(C) Intention. The *Code* is intended to be used by those skilled and knowledgeable in electrical theory, electrical systems, construction, and the installation and operation of electrical equipment. It is not a design specification standard or instruction manual for the untrained and unqualified.

(D) Relation to International Standards. The requirements of the *NEC* address the fundamental safety principles contained in the International Electrotechnical Commission (IEC) standards, including protection against electric shock, adverse thermal effects, overcurrent, fault currents, and overvoltage. **Figure 90–2**

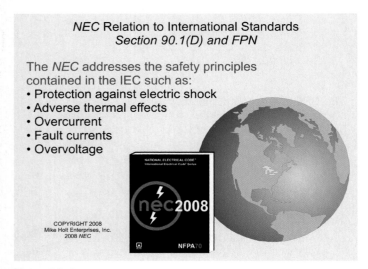

NEC Relation to International Standards
Section 90.1(D) and FPN

The *NEC* addresses the safety principles
contained in the IEC such as:
- Protection against electric shock
- Adverse thermal effects
- Overcurrent
- Fault currents
- Overvoltage

COPYRIGHT 2008
Mike Holt Enterprises, Inc.
2008 *NEC*

NFPA70

Figure 90–2

Author's Comments:

- See the definition of "Overcurrent" in Article 100.

- The *NEC* is used in Chile, Ecuador, Peru, and the Philippines. It's also the electrical code for Colombia, Costa Rica, Mexico, Panama, Puerto Rico, and Venezuela. Because of these adoptions, the *NEC* is available in Spanish from the National Fire Protection Association, 1.617.770.3000.

90.2 Scope of the *NEC*.

(A) What is Covered. The *NEC* contains requirements necessary for the proper electrical installation of electrical conductors, equipment, and raceways; signaling and communications conductors, equipment, and raceways; as well as optical fiber cables and raceways for the following locations: **Figure 90–3**

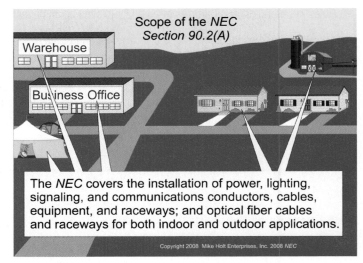

Scope of the *NEC*
Section 90.2(A)

Warehouse

Business Office

The *NEC* covers the installation of power, lighting, signaling, and communications conductors, cables, equipment, and raceways; and optical fiber cables and raceways for both indoor and outdoor applications.

Copyright 2008 Mike Holt Enterprises, Inc. 2008 *NEC*

Figure 90–3

(1) Public and private premises, including buildings or structures, mobile homes, recreational vehicles, and floating buildings.

(2) Yards, lots, parking lots, carnivals, and industrial substations.

(3) Conductors and equipment connected to the utility supply.

(4) Installations used by an electric utility, such as office buildings, warehouses, garages, machine shops, recreational buildings, and other electric utility buildings not

an integral part of a utility's generating plant, substation, or control center. **Figure 90–4**

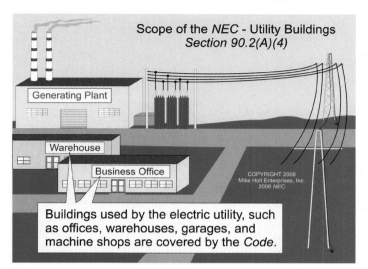

Figure 90–4

(B) What isn't Covered. The *NEC* doesn't apply to:

(1) Transportation Vehicles. Installations in cars, trucks, boats, ships and watercraft, planes, electric trains, or underground mines.

(2) Mining Equipment. Installations underground in mines and self-propelled mobile surface mining machinery and its attendant electrical trailing cables.

(3) Railways. Railway power, signaling, and communications wiring.

(4) Communications Utilities. The installation requirements of the *NEC* don't apply to communications (telephone), CATV, or network-powered broadband utility equipment located in building spaces used exclusively for these purposes, or outdoors if the installation is under the exclusive control of the communications utility. **Figure 90–5**

> **Author's Comment:** Interior wiring for communications systems, not in building spaces used exclusively for these purposes, must be installed in accordance with the following Chapter 8 requirements: **Figure 90–6**
> - Telephone and Data, Article 800
> - CATV, Article 820
> - Network-Powered Broadband, Article 830

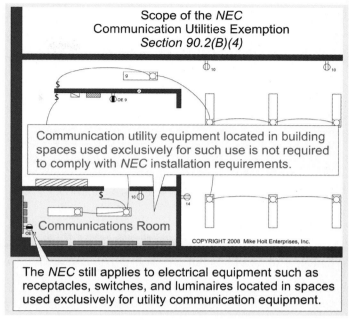

Figure 90–5

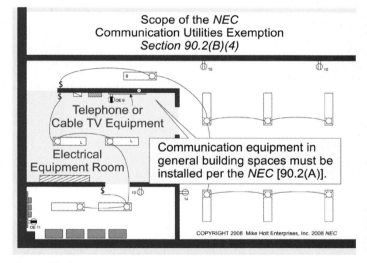

Figure 90–6

(5) Electric Utilities. The *NEC* doesn't apply to installations under the exclusive control of an electric utility where such installations:

    a. Consist of service drops or service laterals and associated metering. **Figure 90–7**

    b. Are located on legally established easements, or rights-of-way recognized by public/utility regulatory agencies, or property owned or leased by the electric utility. **Figure 90–8**

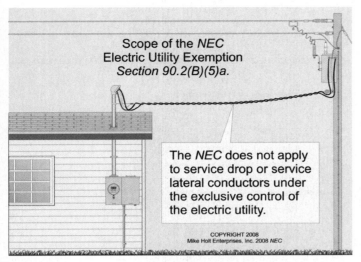

Figure 90–7

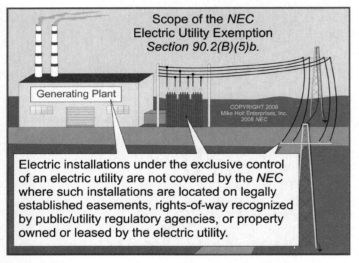

Figure 90–8

c. Are on property owned or leased by the electric utility for the purpose of generation, transformation, transmission, distribution, or metering of electric energy. **Figure 90–9**

Author's Comment: Luminaires located in legally established easements, or rights-of-way, such as at poles supporting transmission or distribution lines, are exempt from the *NEC*. However, if the electric utility provides site and public lighting on private property, then the installation must comply with the *NEC* [90.2(A)(4)].

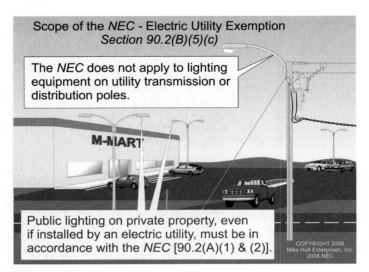

Figure 90–9

FPN to 90.2(B)(4) and (5): Utilities include entities that install, operate, and maintain communications systems (telephone, CATV, Internet, satellite, or data services) or electric supply systems (generation, transmission, or distribution systems) and are designated or recognized by governmental law or regulation by public service/utility commissions. Utilities may be subject to compliance with codes and standards covering their regulated activities as adopted under governmental law or regulation.

90.3 *Code* Arrangement. The *Code* is divided into an Introduction and nine chapters. **Figure 90–10**

General Requirements. The requirements contained in Chapters 1, 2, 3, and 4 apply to all installations.

Special Requirements. The requirements contained in Chapters 5, 6, and 7 apply to special occupancies, special equipment, or other special conditions. They can supplement or modify the requirements in Chapters 1 through 4.

Communications Systems. Chapter 8 contains the requirements for communications systems, such as telephone, antenna wiring, CATV, and network-powered broadband systems. Communications systems aren't subject to the general requirements of Chapters 1 through 4, or the special requirements of Chapters 5 through 7, unless there's a specific reference in Chapter 8 to a rule in Chapters 1 through 7.

Author's Comment: An example of how Chapter 8 works is the rules for working space about equipment. The typical 3 ft working space isn't required in front of communications equipment, because Table 110.26(A)(1) isn't referenced in Chapter 8.

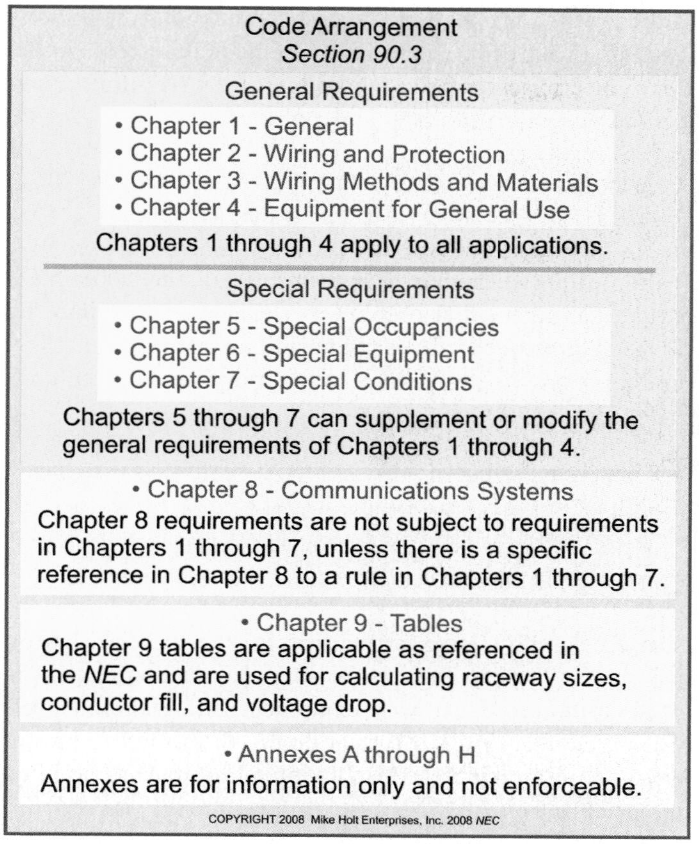

Code Arrangement
Section 90.3

General Requirements

- Chapter 1 - General
- Chapter 2 - Wiring and Protection
- Chapter 3 - Wiring Methods and Materials
- Chapter 4 - Equipment for General Use

Chapters 1 through 4 apply to all applications.

Special Requirements

- Chapter 5 - Special Occupancies
- Chapter 6 - Special Equipment
- Chapter 7 - Special Conditions

Chapters 5 through 7 can supplement or modify the general requirements of Chapters 1 through 4.

- Chapter 8 - Communications Systems

Chapter 8 requirements are not subject to requirements in Chapters 1 through 7, unless there is a specific reference in Chapter 8 to a rule in Chapters 1 through 7.

- Chapter 9 - Tables

Chapter 9 tables are applicable as referenced in the *NEC* and are used for calculating raceway sizes, conductor fill, and voltage drop.

- Annexes A through H

Annexes are for information only and not enforceable.

COPYRIGHT 2008 Mike Holt Enterprises, Inc. 2008 *NEC*

Figure 90–10

Tables. Chapter 9 consists of tables applicable as referenced in the *NEC.* The tables are used to calculate raceway sizing, conductor fill, the radius of conduit and tubing bends, and conductor voltage drop.

Annexes. Annexes aren't part of the *Code*, but are included for informational purposes. There are eight Annexes:

- Annex A. Product Safety Standards
- Annex B. Application Information for Ampacity Calculation
- Annex C. Raceway Fill Tables for Conductors and Fixture Wires of the Same Size
- Annex D. Examples
- Annex E. Types of Construction
- Annex F. Critical Operations Power Systems (COPS)
- Annex G. Supervisory Control and Data Acquisition (SCADA)
- Annex H. Administration and Enforcement

90.4 Enforcement. The *Code* is intended to be suitable for enforcement by governmental bodies that exercise legal jurisdiction over electrical installations for power, lighting, signaling circuits, and communications systems, such as: **Figure 90–11**

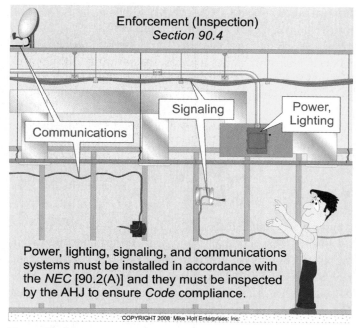

Enforcement (Inspection)
Section 90.4

Communications

Signaling

Power, Lighting

Power, lighting, signaling, and communications systems must be installed in accordance with the *NEC* [90.2(A)] and they must be inspected by the AHJ to ensure *Code* compliance.

COPYRIGHT 2008 Mike Holt Enterprises, Inc.

Figure 90–11

Signaling circuits which include:

- Article 725 Class 1, Class 2, and Class 3 Remote-Control, Signaling, and Power-Limited Circuits
- Article 760 Fire Alarm Systems
- Article 770 Optical Fiber Cables and Raceways

Communications systems which include:

- Article 800 Communications Circuits (twisted-pair conductors)
- Article 810 Radio and Television Equipment (satellite dish and antenna)
- Article 820 Community Antenna Television and Radio Distribution Systems (coaxial cable)
- Article 830 Network-Powered Broadband Communications Systems

Author's Comment: The installation requirements for signaling circuits and communications circuits are covered in Mike Holt's *Understanding the National Electrical Code, Volume 2* textbook.

The enforcement of the *NEC* is the responsibility of the authority having jurisdiction (AHJ), who is responsible for interpreting requirements, approving equipment and materials, waiving *Code* requirements, and ensuring equipment is installed in accordance with listing instructions.

Author's Comment: See the definition of "Authority Having Jurisdiction" in Article 100.

Interpretation of the Requirements. The authority having jurisdiction is responsible for interpreting the *NEC*, but his or her decisions must be based on a specific *Code* requirement. If an installation is rejected, the authority having jurisdiction is legally responsible for informing the installer which specific *NEC* rule was violated.

> **Author's Comment:** The art of getting along with the authority having jurisdiction consists of doing good work and knowing what the *Code* actually says (as opposed to what you only think it says). It's also useful to know how to choose your battles when the inevitable disagreement does occur.

Approval of Equipment and Materials. Only the authority having jurisdiction has authority to approve the installation of equipment and materials. Typically, the authority having jurisdiction will approve equipment listed by a product testing organization, such as Underwriters Laboratories, Inc. (UL), but the *NEC* doesn't require all equipment to be listed. See 90.7, 110.2, 110.3, and the definitions for "Approved," "Identified," "Labeled," and "Listed" in Article 100. **Figure 90–12**

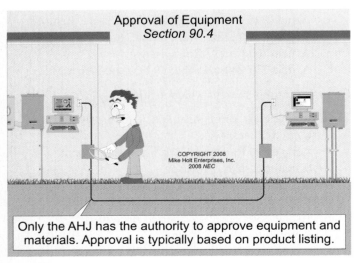

Approval of Equipment
Section 90.4

COPYRIGHT 2008
Mike Holt Enterprises, Inc.
2008 *NEC*

Only the AHJ has the authority to approve equipment and materials. Approval is typically based on product listing.

Figure 90–12

> **Author's Comment:** According to the *NEC*, the authority having jurisdiction determines the approval of equipment. This means he/she can reject an installation of listed equipment and he/she can approve the use of unlisted equipment. Given our highly litigious society, approval of unlisted equipment is becoming increasingly difficult to obtain.

Waiver of Requirements. By special permission, the authority having jurisdiction can waive specific requirements in the *Code* or permit alternative methods where it's assured equivalent safety can be achieved and maintained.

> **Author's Comment:** Special permission is defined in Article 100 as the written consent of the authority having jurisdiction.

Waiver of New Product Requirements. If the 2008 *NEC* requires products that aren't yet available at the time the *Code* is adopted, the authority having jurisdiction can allow products that were acceptable in the previous *Code* to continue to be used.

> **Author's Comment:** Sometimes it takes years before testing laboratories establish product standards for new *NEC* requirements, and then it takes time before manufacturers can design, manufacture, and distribute these products to the marketplace.

Compliance with Listing Instructions. It's the authority having jurisdiction's responsibility to ensure electrical equipment is installed in accordance with equipment listing and/or labeling instructions [110.3(B)]. In addition, the authority having jurisdiction can reject the installation of equipment modified in the field [90.7].

> **Author's Comment:** The *NEC* doesn't address the maintenance of electrical equipment because the *Code* is an installation standard, not a maintenance standard. See NFPA 70B—*Recommended Practice for Electrical Equipment Maintenance.*

90.5 Mandatory Requirements and Explanatory Material.

(A) Mandatory Requirements. In the *NEC* the words "shall" or "shall not," indicate a mandatory requirement.

> **Author's Comment:** For the ease of reading this textbook, the word "shall" has been replaced with the word "must," and the words "shall not" have been replaced with "must not."

(B) Permissive Requirements. When the *Code* uses "shall be permitted" it means the identified actions are allowed but not required, and the authority having jurisdiction is not to restrict an installation from being done in that manner. A permissive rule is often an exception to the general requirement.

> **Author's Comment:** For ease of reading, the phrase "shall be permitted" as used in the *Code*, has been replaced in this textbook with the phrase "is permitted."

(C) Explanatory Material. References to other standards or sections of the *NEC*, or information related to a *Code* rule, are included in the form of Fine Print Notes (FPN). Fine Print Notes are for information only and aren't intended to be enforceable.

For example, Fine Print Note No. 4 in 210.19(A)(1) recommends that the circuit voltage drop not exceed 3 percent. This isn't a requirement; it's just a recommendation.

90.6 Formal Interpretations.
To promote uniformity of interpretation and application of the provisions of the *NEC*, formal interpretation procedures have been established and are found in the NFPA Regulations Governing Committee Projects.

> **Author's Comment:** This is rarely done because it's a very time-consuming process, and formal interpretations from the NFPA are not binding on the authority having jurisdiction.

90.7 Examination of Equipment for Product Safety.
Product evaluation for safety is typically performed by a testing laboratory, which publishes a list of equipment that meets a nationally recognized test standard. Products and materials listed, labeled, or identified by a testing laboratory are generally approved by the authority having jurisdiction.

> **Author's Comment:** See Article 100 for the definition of "Approved."

Listed, factory-installed, internal wiring and construction of equipment need not be inspected at the time of installation, except to detect alterations or damage [300.1(B)]. Figure 90–13

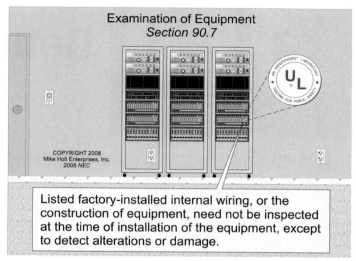

Examination of Equipment
Section 90.7

COPYRIGHT 2008
Mike Holt Enterprises, Inc.
2008 *NEC*

Listed factory-installed internal wiring, or the construction of equipment, need not be inspected at the time of installation of the equipment, except to detect alterations or damage.

Figure 90–13

90.9 Units of Measurement.

(B) Dual Systems of Units. Both the metric and inch-pound measurement systems are shown in the *NEC*, with the metric units appearing first and the inch-pound system immediately following in parentheses.

> **Author's Comment:** This is the standard practice in all NFPA standards, even though the U.S. construction industry uses inch-pound units of measurement.

(D) Compliance. Installing electrical systems in accordance with the metric system or the inch-pound system is considered to comply with the *Code*.

> **Author's Comment:** Since compliance with either the metric or the inch-pound system of measurement constitutes compliance with the *NEC*, this textbook uses only inch-pound units.

100 Definitions

INTRODUCTION TO ARTICLE 100—DEFINITIONS

Have you ever had a conversation with someone, only to discover what you said and what he/she heard were completely different? This often happens when people in a conversation don't understand the definitions of the words being used, and that's why the definitions of key terms are located right at the beginning of the *NEC* (Article 100), or the beginning of each Article.

If we can all agree on important definitions, then we speak the same language and avoid misunderstandings. Because the *Code* exists to protect people and property, we can agree it's very important to know the definitions presented in Article 100.

Now, here are a couple of things you may not know about Article 100:

Article 100 contains the definitions of many, but not all, of the terms used throughout the *NEC*. In general, only those terms used in two or more articles are defined in Article 100.

- Part I of Article 100 contains the definitions of terms used throughout the *Code*.

- Part II of Article 100 contains only terms that apply to systems that operate at over 600V nominal.

How can you possibly learn all of these definitions? There seem to be so many. Here are a few tips:

- Break the task down. Study a few words at a time, rather than trying to learn them all at one sitting.

- Review the graphics in the textbook. These will help you see how a term is applied.

- Relate them to your work. As you read a word, think about how it applies to the work you're doing. This will provide a natural reinforcement to the learning process.

DEFINITIONS

Authority Having Jurisdiction (AHJ). The organization, office, or individual responsible for approving equipment, materials, an installation, or a procedure. See 90.4 and 90.7 for more information.

> **FPN:** The authority having jurisdiction may be a federal, state, or local government, or an individual such as a fire chief, fire marshal, chief of a fire prevention bureau or labor department or health department, a building official or electrical inspector, or others having statutory authority. In some circumstances, the property owner or his/her agent assumes the role, and at government installations, the commanding officer, or departmental official may be the authority having jurisdiction.

Author's Comments:

- Typically, the authority having jurisdiction is the electrical inspector who has legal statutory authority. In the absence of federal, state, or local regulations, the operator of the facility or his/her agent, such as an architect or engineer of the facility, can assume the role.

- Some believe the authority having jurisdiction should have a strong background in the electrical field, such as having studied electrical engineering or having obtained an electrical contractor's license, and in a few states this is a legal requirement. Memberships, certifications, and active participation in electrical organizations, such as the IAEI (www.IAEI.org), speak to an individual's qualifications.

Bonded (Bonding). Connected to establish electrical continuity and conductivity. **Figure 100–1**

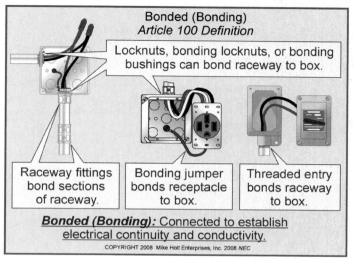

Bonded (Bonding)
Article 100 Definition

Locknuts, bonding locknuts, or bonding bushings can bond raceway to box.

Raceway fittings bond sections of raceway.

Bonding jumper bonds receptacle to box.

Threaded entry bonds raceway to box.

Bonded (Bonding): Connected to establish electrical continuity and conductivity.

COPYRIGHT 2008 Mike Holt Enterprises, Inc. 2008 *NEC*

Figure 100–1

Author's Comment: The purpose of bonding is to connect two or more conductive objects together to ensure the electrical continuity of the fault current path, provide the capacity and ability to conduct safely any fault current likely to be imposed, and to minimize potential differences (voltage) between conductive components. **Figure 100–2**

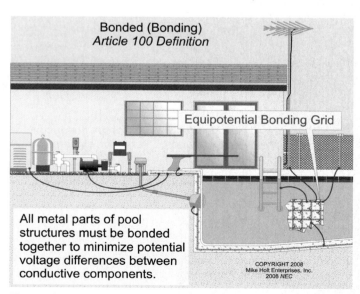

Bonded (Bonding)
Article 100 Definition

Equipotential Bonding Grid

All metal parts of pool structures must be bonded together to minimize potential voltage differences between conductive components.

COPYRIGHT 2008 Mike Holt Enterprises, Inc. 2008 *NEC*

Figure 100–2

Bonding Jumper. A conductor sized to ensure electrical conductivity between metal parts of the electrical installation. **Figure 100–3**

Bonding Jumper
Article 100 Definition

COPYRIGHT 2008 Mike Holt Enterprises, Inc. 2008 *NEC*

Bonding Jumper: A conductor used to ensure electrical conductivity between metal parts that must be electrically connected.

Figure 100–3

Bonding Jumper, Main. A conductor, screw, or strap that connects the circuit equipment grounding conductor to the neutral conductor at service equipment in accordance with 250.24(B) [250.24(A)(4), 250.28, and 408.3(C)]. **Figure 100–4**

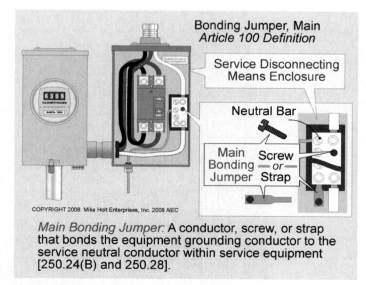

Bonding Jumper, Main
Article 100 Definition

Service Disconnecting Means Enclosure

Neutral Bar

Main Bonding Jumper

Screw —or— Strap

COPYRIGHT 2008 Mike Holt Enterprises, Inc. 2008 *NEC*

Main Bonding Jumper: A conductor, screw, or strap that bonds the equipment grounding conductor to the service neutral conductor within service equipment [250.24(B) and 250.28].

Figure 100–4

Ground. The earth. **Figure 100–5**

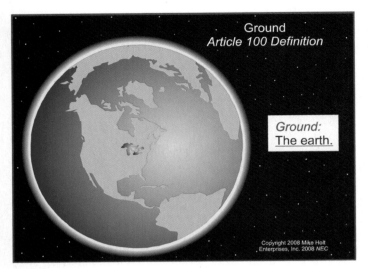

Figure 100–5

Grounded (Grounding). Connected to ground or to a conductive body that extends the ground connection.

> **Author's Comment:** An example of a "body that extends the ground (earth) connection" is the termination to structural steel that is connected to the earth either directly or by the termination to another grounding electrode. **Figure 100–6**

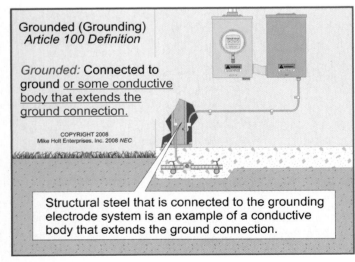

Figure 100–6

Grounded, Solidly. Connected to ground without inserting any resistor or impedance device. **Figure 100–7**

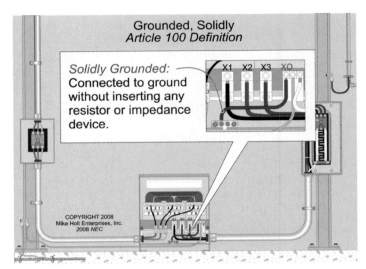

Figure 100–7

Grounded Conductor. The conductor that is intentionally grounded (connected to the earth). **Figure 100–8**

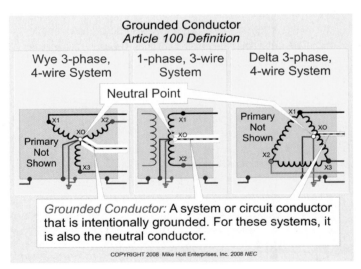

Figure 100–8

> **Author's Comment:** Because the neutral conductor of a solidly grounded system is always grounded (connected to the earth), it's both a "grounded conductor" and a "neutral" conductor. To make it easier for the reader of this textbook, we will refer to the "grounded" conductor of a solidly grounded system as the "neutral" conductor.

Grounding Conductor. A conductor used to connect equipment or the neutral conductor of a wiring system to a grounding electrode. **Figure 100–9**

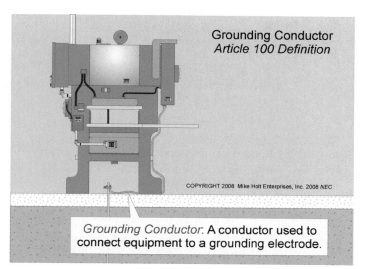

Grounding Conductor
Article 100 Definition

COPYRIGHT 2008 Mike Holt Enterprises, Inc. 2008 *NEC*

Grounding Conductor. A conductor used to connect equipment to a grounding electrode.

Figure 100–9

Author's Comment: The neutral conductor is only permitted to be grounded (connected to the earth) at specific locations as identified in 250.142.

Grounding Conductor, Equipment (EGC). The conductive path that connects metal parts of equipment to the system neutral conductor or to the grounding electrode conductor, or both [250.110 through 250.126]. **Figure 100–10**

FPN No. 1: The circuit equipment grounding conductor also performs bonding.

Author's Comment: To quickly remove dangerous touch voltage on metal parts from a ground fault, the equipment grounding conductor must have sufficiently low impedance to the source so that fault current will quickly rise to a level that will open the branch-circuit overcurrent device [250.2 and 250.4(A)(3)].

FPN No. 2: An equipment grounding conductor can be any one or a combination of the types listed in 250.118. **Figure 100–11**

Author's Comment: Equipment grounding conductors include:

- Bare or insulated conductor
- Rigid Metal Conduit
- Intermediate Metal Conduit
- Electrical Metallic Tubing
- Listed Flexible Metal Conduit as limited by 250.118(5)

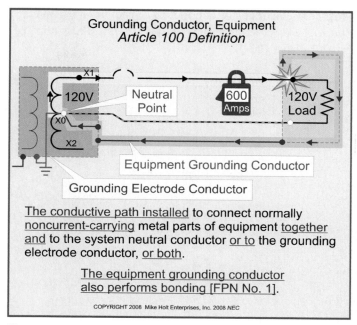

Grounding Conductor, Equipment
Article 100 Definition

120V — Neutral Point — 600 Amps — 120V Load

X1
X0
X2

Equipment Grounding Conductor

Grounding Electrode Conductor

The conductive path installed to connect normally noncurrent-carrying metal parts of equipment together and to the system neutral conductor or to the grounding electrode conductor, or both.

The equipment grounding conductor also performs bonding [FPN No. 1].

COPYRIGHT 2008 Mike Holt Enterprises, Inc. 2008 *NEC*

Figure 100–10

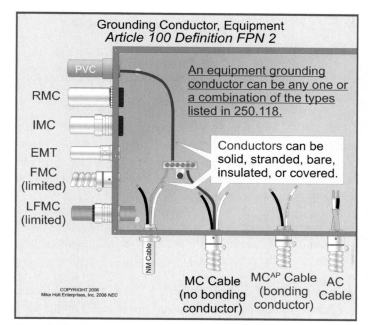

Grounding Conductor, Equipment
Article 100 Definition FPN 2

PVC
RMC
IMC
EMT
FMC (limited)
LFMC (limited)

An equipment grounding conductor can be any one or a combination of the types listed in 250.118.

Conductors can be solid, stranded, bare, insulated, or covered.

NM Cable

MC Cable (no bonding conductor)

MC^AP Cable (bonding conductor)

AC Cable

COPYRIGHT 2008 Mike Holt Enterprises, Inc. 2008 *NEC*

Figure 100–11

- Listed Liquidtight Flexible Metal Conduit as limited by 250.118(6)
- Armored Cable
- Copper metal sheath of Mineral Insulated Cable
- Metal Clad Cable as limited by 250.118(10)
- Metallic cable trays as limited by 250.118(11) and 392.60
- Electrically continuous metal raceways listed for grounding
- Surface Metal Raceways listed for grounding

Grounding Electrode. A conducting object used to make a direct electrical connection to the earth [250.50 through 250.70]. **Figure 100–12**

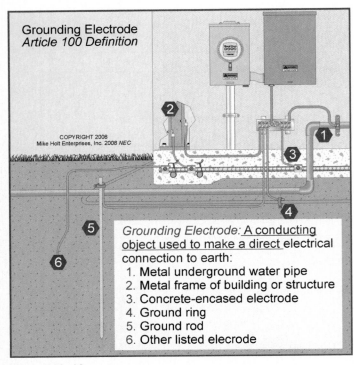

Figure 100–12

Grounding Electrode Conductor. The conductor used to connect the system neutral conductor or the metal parts of electrical equipment to a grounding electrode or to a point on the grounding electrode system. **Figure 100–13**

Author's Comment: For services see 250.24(A), for separately derived systems see 250.30(A), and buildings or structures supplied by a feeder see 250.32(A).

Intersystem Bonding Termination. A device that provides a means to connect communications systems grounding and bonding conductors to the building grounding electrode system at the service equipment or at the disconnecting means for buildings or structures supplied by a feeder in accordance with 250.94. **Figure 100–14**

Neutral Conductor. The conductor connected to the neutral point of a system that is intended to carry current under normal conditions. **Figure 100–15**

Author's Comment: The neutral conductor of a solidly grounded system is required to be grounded (connected to the earth), therefore this conductor is also called a "grounded conductor."

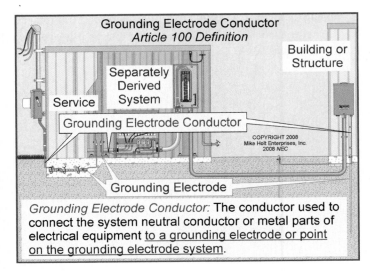

Figure 100–13

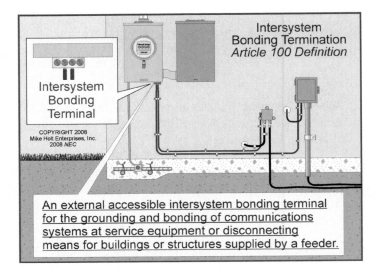

Figure 100–14

Neutral Point. The common point of a 4-wire, three-phase, wye-connected system; the midpoint of a 3-wire, single-phase system; or the midpoint of the single-phase portion of a three-phase, delta-connected system. **Figure 100–16**

Ungrounded. Not connected to the ground (earth) or a conductive body that extends the ground (earth) connection. **Figure 100–17**

Author's Comment: The use of this term relates to an ungrounded system, where the system windings are not grounded (connected to the earth) [250.4(B) and 250.30(B)].

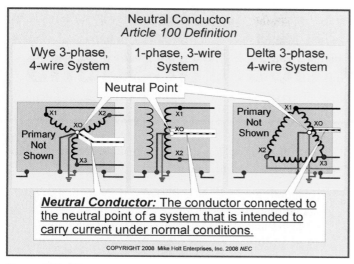

Figure 100–15

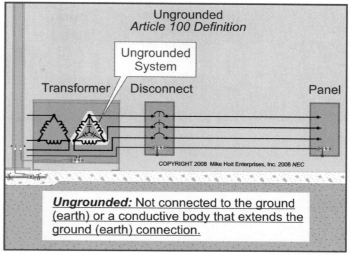

Figure 100–17

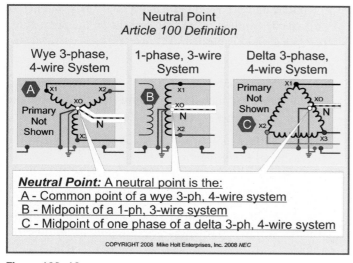

Figure 100–16

ARTICLES 90–100 Practice Questions

Use the 2008 *NEC* to answer the following questions.

ARTICLES 90 THROUGH 100 PRACTICE QUESTIONS

1. The *NEC* is _____.

    (a) intended to be a design manual
    (b) meant to be used as an instruction guide for untrained persons
    (c) for the practical safeguarding of persons and property
    (d) published by the Bureau of Standards

2. Compliance with the provisions of the *NEC* will result in _____.

    (a) good electrical service
    (b) an efficient electrical system
    (c) an electrical system essentially free from hazard
    (d) all of these

3. The *Code* contains provisions considered necessary for safety, which will not necessarily result in _____ .

    (a) efficient use
    (b) convenience
    (c) good service or future expansion of electrical use
    (d) all of these

4. Hazards often occur because of _____.

    (a) overloading of wiring systems by methods or usage not in conformity with the *NEC*
    (b) initial wiring not providing for increases in the use of electricity
    (c) a and b
    (d) none of these

5. The *Code* isn't a design specification standard or instruction manual for the untrained and unqualified.

    (a) True
    (b) False

6. The following systems shall be installed in accordance with the *NEC* requirements:

    (a) signaling
    (b) communications
    (c) electrical conductors, equipment, and raceways
    (d) all of these

7. The *NEC* applies to the installation of _____.

    (a) electrical conductors and equipment within or on public and private buildings
    (b) outside conductors and equipment on the premises
    (c) optical fiber cables
    (d) all of these

8. This *Code* covers the installation of _____ for public and private premises, including buildings, structures, mobile homes, recreational vehicles, and floating buildings.

    (a) optical fiber cables
    (b) electrical equipment
    (c) raceways
    (d) all of these

9. The *NEC* does not cover electrical installations in ships, watercraft, railway rolling stock, aircraft, or automotive vehicles.

    (a) True
    (b) False

10. The *Code* covers underground mine installations and self-propelled mobile surface mining machinery and its attendant electrical trailing cable.

    (a) True
    (b) False

11. Installations of communications equipment that are under the exclusive control of communications utilities, and located outdoors or in building spaces used exclusively for such installations _____ covered by the *NEC*.

    (a) are
    (b) are sometimes
    (c) are not
    (d) maybe

12. Electric utilities may include entities that install, operate, and maintain _____.

    (a) communications systems (telephone, CATV, Internet, satellite, or data services)
    (b) electric supply systems (generation, transmission, or distribution systems)
    (c) local area network wiring on the premises
    (d) a or b

13. Utilities may be subject to compliance with codes and standards covering their regulated activities as adopted under governmental law or regulation.

    (a) True
    (b) False

14. Utilities may include entities that are designated or recognized by governmental law or regulation by public service/utility commissions.

    (a) True
    (b) False

15. The *NEC* does not apply to electric utility-owned wiring and equipment _____.

    (a) installed by an electrical contractor
    (b) installed on public property
    (c) consisting of service drops installed by a utility
    (d) in a utility office building

16. Chapters 1 through 4 of the *NEC* apply _____.

    (a) generally to all electrical installations
    (b) only to special occupancies and conditions
    (c) only to special equipment and material
    (d) all of these

17. Communications wiring such as telephone, antenna, and CATV wiring within a building shall not be required to comply with the installation requirements of Chapters 1 through 7, except where it's specifically referenced in Chapter 8.

    (a) True
    (b) False

18. The material located in the *NEC* Annexes are part of the *Code* and shall be complied with.

    (a) True
    (b) False

19. The authority having jurisdiction shall not be required to enforce any requirements of Chapter 7 (Special Conditions) or Chapter 8 (Communications Systems).

    (a) True
    (b) False

20. The _____ has the responsibility for deciding on the approval of equipment and materials.

    (a) manufacturer
    (b) authority having jurisdiction
    (c) testing agency
    (d) none of these

21. By special permission, the authority having jurisdiction may waive specific requirements in this *Code* where it's assured that equivalent objectives can be achieved by establishing and maintaining effective safety.

    (a) True
    (b) False

22. The authority having jurisdiction has the responsibility _____.

    (a) for making interpretations of rules
    (b) for deciding upon the approval of equipment and materials
    (c) for waiving specific requirements in the *Code* and permitting alternate methods and material if safety is maintained
    (d) all of these

23. If the *NEC* requires new products that are not yet available at the time a new edition is adopted, the _____ may permit the use of the products that comply with the previous edition of the *Code* adopted by that jurisdiction.

(a) electrical engineer
(b) master electrician
(c) authority having jurisdiction
(d) permit holder

24. In the *NEC,* the words _____ indicate a mandatory requirement.

(a) shall
(b) shall not
(c) shall be permitted
(d) a or b

25. When the *Code* uses ____, it means the identified actions are allowed but not required, and the authority having jurisdiction is not to restrict an installation from being done in that manner.

(a) shall
(b) shall not
(c) shall be permitted
(d) a or b

26. Explanatory material, such as references to other standards, references to related sections of the *NEC*, or information related to a *Code* rule, are included in the form of Fine Print Notes (FPNs).

(a) True
(b) False

27. Factory-installed _____ wiring of equipment need not be inspected at the time of installation of the equipment, except to detect alterations or damage.

(a) external
(b) associated
(c) internal
(d) all of these

28. Compliance with either the metric or the inch-pound unit of measurement system shall be permitted.

(a) True
(b) False

29. Acceptable to the authority having jurisdiction means _____.

(a) identified
(b) listed
(c) approved
(d) labeled

30. Where no statutory requirement exists, the authority having jurisdiction can be a property owner or his/her agent, such as an architect or engineer.

(a) True
(b) False

31. Bonded can be described as _____ to establish electrical continuity and conductivity.

(a) isolated
(b) guarded
(c) connected
(d) separated

32. The connection between the grounded conductor and the equipment grounding conductor at the service is accomplished by installing a(n) _____ jumper.

(a) main bonding
(b) system bonding
(c) equipment bonding
(d) circuit bonding

33. The word "Earth" best describes what *NEC* term?

(a) Bonded
(b) Ground
(c) Effective ground-fault current path
(d) Guarded

34. "Connected to ground or to a conductive body that extends the ground connection" is called _____.

(a) equipment grounding
(b) bonded
(c) grounded
(d) all of these

35. "Connected to ground without the insertion of any resistor or impedance device" is _____.

(a) grounded
(b) solidly grounded
(c) effectively grounded
(d) grounding conductor

36. A circuit conductor that is intentionally grounded is called a(n) _____.

    (a) grounding conductor
    (b) unidentified conductor
    (c) grounded conductor
    (d) grounding electrode conductor

37. The installed conductive path that connects normally noncurrent-carrying metal parts of equipment together and to the system grounded conductor or to the grounding electrode conductor, or both, is known as a(n) _____.

    (a) grounding electrode conductor
    (b) grounding conductor
    (c) equipment grounding conductor
    (d) none of these

38. A conducting object through which a direct connection to earth is established is a _____.

    (a) bonding conductor
    (b) grounding conductor
    (c) grounding electrode
    (d) grounded conductor

39. A conductor used to connect the system grounded conductor or equipment to a grounding electrode or to a point on the grounding electrode system is called the _____ conductor.

    (a) main grounding
    (b) common main
    (c) equipment grounding
    (d) grounding electrode

40. A device that provides a means to connect _____ systems grounding and bonding conductors to the building grounding electrode system is an intersystem bonding termination.

    (a) limited energy
    (b) low-voltage
    (c) communications
    (d) power and lighting

41. A neutral conductor is the conductor connected to the _____ of a system, which is intended to carry neutral current under normal conditions.

    (a) grounding electrode
    (b) neutral point
    (c) intersystem bonding termination
    (d) none of these

42. The common point on a wye-connection in a polyphase system describes a neutral point.

    (a) True
    (b) False

43. Ungrounded describes "not connected to ground or a conductive body that extends the ground connection."

    (a) True
    (b) False

Notes

250 Grounding and Bonding

INTRODUCTION TO ARTICLE 250—GROUNDING AND BONDING

No other article can match Article 250 for misapplication, violation, and misinterpretation. People often insist on building electrical installations that violate this article.

Article 250 covers the grounding requirements for providing a low-impedance path to the earth to reduce overvoltage from lightning, and the bonding requirements for a low-impedance fault current path necessary to facilitate the operation of overcurrent devices in the event of a ground fault.

Over the past four *Code* cycles, this article was extensively revised to organize it better and make it easier to understand and implement. It's arranged in a logical manner, so it's a good idea to just read through Article 250 to get a big picture view—after you review the definitions. Next, study the article closely so you understand the details. The illustrations will help you understand the key points.

250 Grounding and Bonding— Part I. General

ARTICLE

PART I. GENERAL

250.1 Scope. Article 250 contains the following grounding and bonding requirements:

(1) What systems and equipment are required to be grounded

(3) Location of grounding connections

(4) Types of electrodes and sizes of grounding and bonding conductors

(5) Methods of grounding and bonding

250.2 Definitions.

Bonding Jumper, System. The conductor, screw, or strap that bonds the metal parts of a separately derived system to the system neutral point according to 250.30(A)(1).**Figure 250–1**

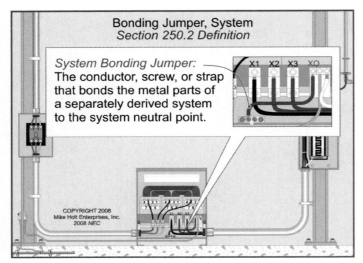

Figure 250–1

Author's Comment: The system bonding jumper provides a low-impedance fault current path to the power supply to facilitate the clearing of a ground fault by opening the circuit overcurrent device. For more information, see 250.4(A)(5), 250.28, and 250.30(A)(1).

Effective Ground-Fault Current Path. An intentionally constructed low-impedance conductive path designed to carry fault current from the point of a ground fault on a wiring system to the electrical supply source. **Figure 250–2**

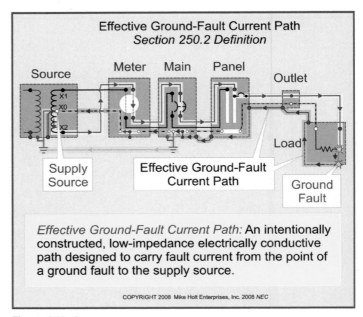

Figure 250–2

The effective ground-fault current path is intended to help remove dangerous voltage from a ground fault by opening the circuit overcurrent device. **Figure 250–3**

Ground Fault. An unintentional connection between an ungrounded conductor and the metal parts of enclosures, raceways, or equipment. **Figure 250–4**

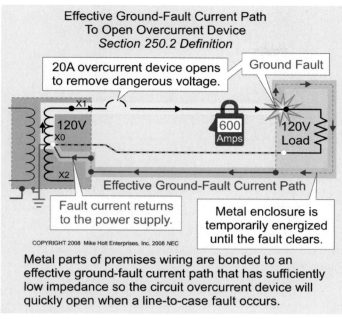

Effective Ground-Fault Current Path
To Open Overcurrent Device
Section 250.2 Definition

Figure 250–3

Metal parts of premises wiring are bonded to an effective ground-fault current path that has sufficiently low impedance so the circuit overcurrent device will quickly open when a line-to-case fault occurs.

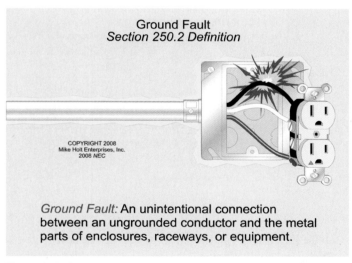

Ground Fault
Section 250.2 Definition

Ground Fault: An unintentional connection between an ungrounded conductor and the metal parts of enclosures, raceways, or equipment.

Figure 250–4

Ground-Fault Current Path. An electrically conductive path from a ground fault to the electrical supply source.

> **FPN:** The ground-fault current path could be metal raceways, cable sheaths, electrical equipment, or other electrically conductive materials, such as metallic water or gas piping, steel-framing members, metal ducting, reinforcing steel, or the shields of communications cables. **Figure 250–5**

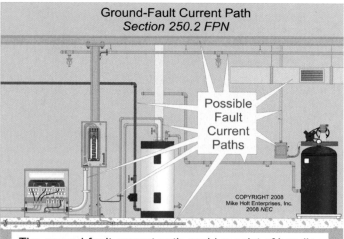

Ground-Fault Current Path
Section 250.2 FPN

The ground-fault current path could consist of bonding conductors, metal raceways, metal cable sheaths, metal enclosures, and other electrically conductive materials, such as metallic water or gas piping, steel-framing members, metal ducting, reinforcing steel, or the shields of communications cables.

Figure 250–5

Author's Comment: The difference between an "effective ground-fault current path" and a "fault current path" is the effective ground-fault current path is "intentionally" constructed to provide a low-impedance fault current path to the electrical supply source for the purpose of clearing a ground fault. A ground-fault current path is all of the available conductive paths over which fault current flows on its return to the electrical supply source during a ground fault.

250.4 General Requirements for Grounding and Bonding.

(A) Solidly Grounded Systems.

(1) Electrical System Grounding. Electrical power systems, such as the secondary winding of a transformer are grounded (connected to the earth) to limit the voltage caused by lightning, line surges, or unintentional contact by higher-voltage lines. **Figure 250–6**

Author's Comment: System grounding helps reduce fires in buildings as well as voltage stress on electrical insulation, thereby ensuring longer insulation life for motors, transformers, and other system components. **Figure 250–7**

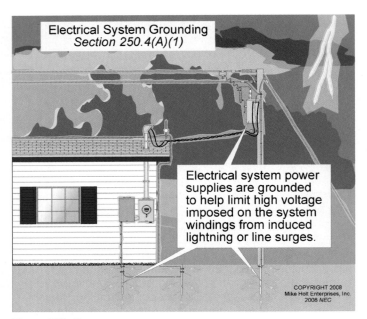

Figure 250–6

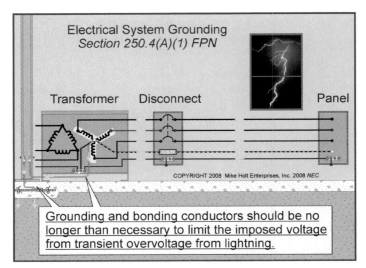

Figure 250–8

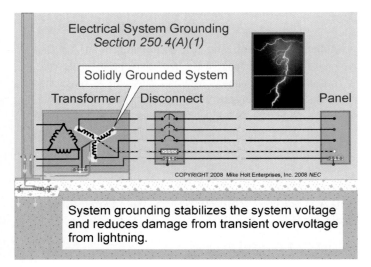

Figure 250–7

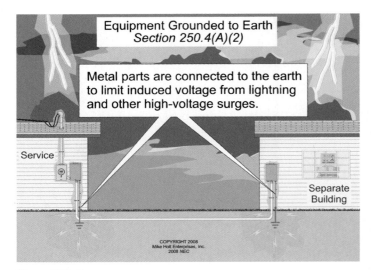

Figure 250–9

FPN: An important consideration for limiting the imposed voltage is the routing of bonding and grounding conductors so that they are not any longer than necessary to complete the connection without disturbing the permanent parts of the installation and so that unnecessary bends and loops are avoided. **Figure 250–8**

(2) Equipment Grounding. Metal parts of electrical equipment are grounded (connected to the earth) to reduce induced voltage on metal parts from exterior lightning so as to prevent fires from an arc within the building or structure. **Figure 250–9**

DANGER: *Failure to ground the metal parts can result in high-voltage on metal parts from an indirect lightning strike to seek a path to the earth within the building—possibly resulting in a fire and or electric shock.* **Figure 250-10**

Author's Comment: Grounding metal parts helps drain off static electricity charges before flashover potential is reached. Static grounding is often used in areas where the discharge (arcing) of the voltage buildup (static) can cause dangerous or undesirable conditions [500.4 FPN No. 3].

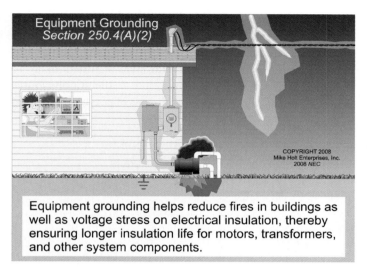

Equipment Grounding
Section 250.4(A)(2)

Equipment grounding helps reduce fires in buildings as well as voltage stress on electrical insulation, thereby ensuring longer insulation life for motors, transformers, and other system components.

Figure 250–10

DANGER: *Because the contact resistance of an electrode to the earth is so high, very little fault current returns to the power supply if the earth is the only fault current return path. Result—the circuit overcurrent device will not open and clear the ground fault, and all metal parts associated with the electrical installation, metal piping, and structural building steel will become and remain energized* **Figure 250–11**

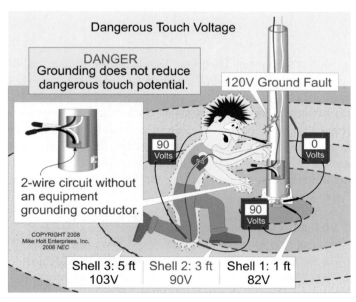

Dangerous Touch Voltage

DANGER
Grounding does not reduce dangerous touch potential.

120V Ground Fault

90 Volts   0 Volts

2-wire circuit without an equipment grounding conductor.

90 Volts

| Shell 3: 5 ft | Shell 2: 3 ft | Shell 1: 1 ft |
| 103V | 90V | 82V |

Figure 250–11

(3) Equipment Bonding. Metal parts of electrical raceways, cables, enclosures, and equipment must be connected to the supply source via an equipment grounding conductor of a type recognized in 250.118. **Figures 250–12** *and* **250–13**

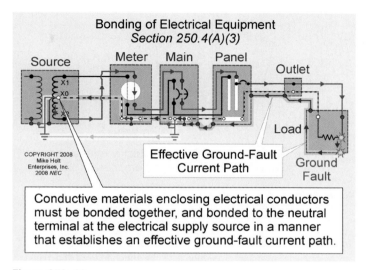

Bonding of Electrical Equipment
Section 250.4(A)(3)

Source   Meter   Main   Panel   Outlet

Effective Ground-Fault Current Path

Load   Ground Fault

Conductive materials enclosing electrical conductors must be bonded together, and bonded to the neutral terminal at the electrical supply source in a manner that establishes an effective ground-fault current path.

Figure 250–12

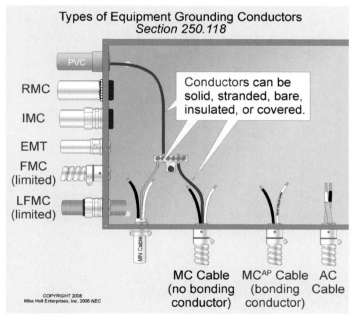

Types of Equipment Grounding Conductors
Section 250.118

PVC
RMC
IMC
EMT
FMC (limited)
LFMC (limited)

Conductors can be solid, stranded, bare, insulated, or covered.

MC Cable (no bonding conductor)   MC^AP Cable (bonding conductor)   AC Cable

Figure 250–13

Author's Comment: To quickly remove dangerous touch voltage on metal parts from a ground fault, the fault current path must have sufficiently low impedance to the source so that fault current will quickly rise to a level that will open the branch-circuit overcurrent device. **Figure 250–14**

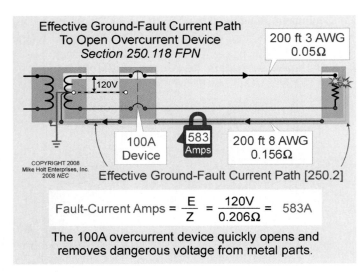

Figure 250–14

Effective Ground-Fault Current Path
To Open Overcurrent Device
Section 250.118 FPN

200 ft 3 AWG
0.05Ω

120V

100A Device

583 Amps

200 ft 8 AWG
0.156Ω

COPYRIGHT 2008
Mike Holt Enterprises, Inc.
2008 NEC

Effective Ground-Fault Current Path [250.2]

$$\text{Fault-Current Amps} = \frac{E}{Z} = \frac{120V}{0.206\Omega} = 583A$$

The 100A overcurrent device quickly opens and removes dangerous voltage from metal parts.

Author's Comment: The time it takes for an overcurrent device to open is inversely proportional to the magnitude of the fault current. This means the higher the ground-fault current value, the less time it will take for the overcurrent device to open and clear the fault. For example, a 20A circuit with an overload of 40A (two times the 20A rating) takes 25 to 150 seconds to open the overcurrent device. At 100A (five times the 20A rating) the 20A breaker trips in 5 to 20 seconds. **Figure 250–15**

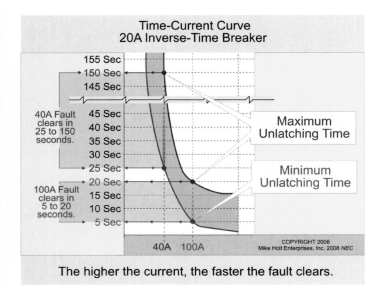

Time-Current Curve
20A Inverse-Time Breaker

155 Sec
150 Sec
145 Sec

40A Fault clears in 25 to 150 seconds.

45 Sec
40 Sec
35 Sec
30 Sec
25 Sec
20 Sec

100A Fault clears in 5 to 20 seconds.

15 Sec
10 Sec
5 Sec

Maximum Unlatching Time

Minimum Unlatching Time

40A 100A

COPYRIGHT 2008
Mike Holt Enterprises, Inc. 2008 NEC

The higher the current, the faster the fault clears.

Figure 250–15

(4) Bonding Conductive Materials. Electrically conductive materials such as metal water piping systems, metal sprinkler piping, metal gas piping, and other metal-piping systems, as well as exposed structural steel members likely to become energized, must be connected to the supply source via an equipment grounding conductor of a type recognized in 250.118. **Figure 250–16**

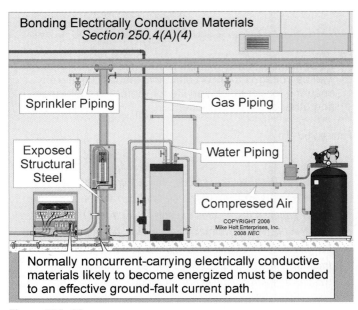

Bonding Electrically Conductive Materials
Section 250.4(A)(4)

Sprinkler Piping

Gas Piping

Exposed Structural Steel

Water Piping

Compressed Air

COPYRIGHT 2008
Mike Holt Enterprises, Inc.
2008 NEC

Normally noncurrent-carrying electrically conductive materials likely to become energized must be bonded to an effective ground-fault current path.

Figure 250–16

Author's Comment: The phrase "likely to become energized" is subject to interpretation by the authority having jurisdiction.

(5) Effective Ground-Fault Current Path. Metal parts of electrical raceways, cables, enclosures, or equipment must be bonded together and to the supply system in a manner that creates a low-impedance path for ground-fault current that facilitates the operation of the circuit overcurrent device. **Figure 250–17**

Author's Comment: To assure a low-impedance ground-fault current path, all circuit conductors must be grouped together in the same raceway, cable, or trench [300.3(B), 300.5(I), and 300.20(A)]. **Figure 250–18**

Because the earth is not suitable to serve as the required effective ground-fault current path, an equipment grounding conductor is required to be run with all circuits. **Figure 250–19**

Effective Ground-Fault Current Path
Section 250.4(A)(5)

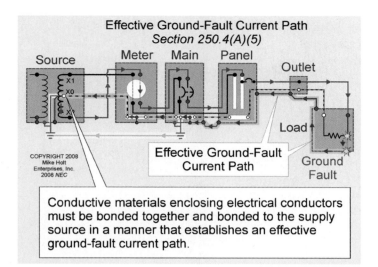

Conductive materials enclosing electrical conductors must be bonded together and bonded to the supply source in a manner that establishes an effective ground-fault current path.

Figure 250–17

All Conductors Must be Grouped
Section 300.3(B)

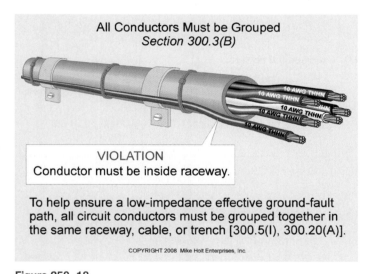

VIOLATION
Conductor must be inside raceway.

To help ensure a low-impedance effective ground-fault path, all circuit conductors must be grouped together in the same raceway, cable, or trench [300.5(I), 300.20(A)].

COPYRIGHT 2008 Mike Holt Enterprises, Inc.

Figure 250–18

Effective Ground-Fault Current Path
Section 250.4(A)(5)

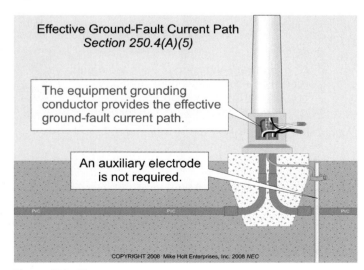

The equipment grounding conductor provides the effective ground-fault current path.

An auxiliary electrode is not required.

COPYRIGHT 2008 Mike Holt Enterprises, Inc. 2008 NEC

Figure 250–19

Question: What is the maximum fault current that can flow through the earth to the power supply from a 120V ground fault to metal parts of a light pole that is grounded (connected to the earth) via a ground rod having a contact resistance to the earth of 25 ohms? **Figure 250–20**

(a) 4.80A   (b) 20A   (c) 40A   (d) 100A

Answer: (a) 4.80A

$I = E/R$
$I = 120V/25$ ohms
$I = 4.80A$

Figure 250–20

DANGER: Because the contact resistance of an electrode to the earth is so high, very little fault current returns to the power supply if the earth is the only fault current return path. Result—the circuit overcurrent device will not open and all metal parts associated with the electrical installation, metal piping, and structural building steel will become and remain energized. **Figure 250–21**

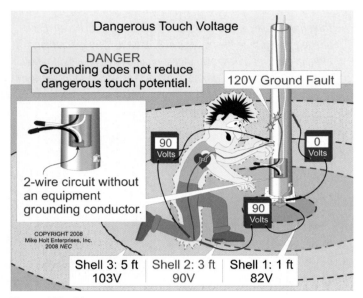

Figure 250–21

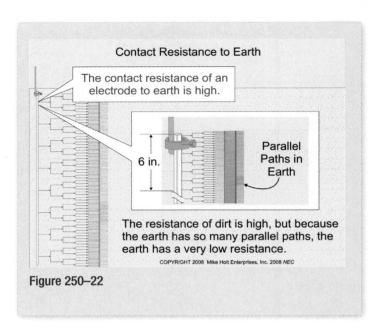

Figure 250–22

SPECIAL SECTION

According to ANSI/IEEE 142, *Recommended Practice for Grounding of Industrial and Commercial Power Systems* (Green Book) [4.1.1], the resistance of the soil outward from a ground rod is equal to the sum of the series resistances of the earth shells. The shell nearest the rod has the highest resistance and each successive shell has progressively larger areas and progressively lower resistances. Don't be concerned if you don't understand this statement; just review the table below. Figure 250–22

Distance from Rod	Soil Contact Resistance
1 ft (Shell 1)	68% of total contact resistance
3 ft (Shells 1 and 2)	75% of total contact resistance
5 ft (Shells 1, 2, and 3)	86% of total contact resistance

Since voltage is directly proportional to resistance, the voltage gradient of the earth around an energized ground rod will be as follows, assuming a 120V ground fault:

Distance from Rod	Soil Contact Resistance	Voltage Gradient
1 ft (Shell 1)	68%	82V
3 ft (Shells 1 and 2)	75%	90V
5 ft (Shells 1, 2, and 3)	86%	103V

(B) Ungrounded Systems.

Author's Comment: Ungrounded systems are those systems with no connection to the ground or a conductive body that extends the ground connection [Article 100]. **Figure 250–23**

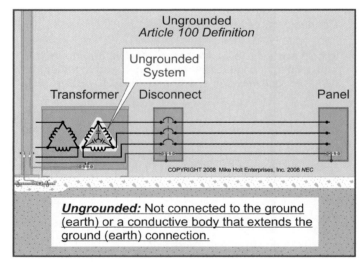

Figure 250–23

(1) Equipment Grounding. Metal parts of electrical equipment are grounded (connected to the earth) to reduce induced voltage on metal parts from exterior lightning so as to prevent fires from an arc within the building or structure. **Figure 250–24**

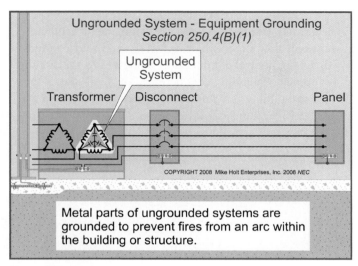

Figure 250–24

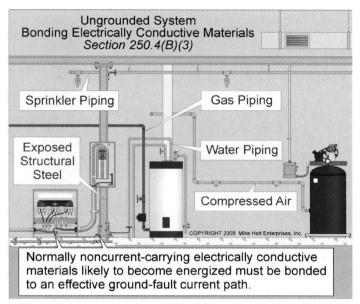

Figure 250–25

Author's Comment: Grounding metal parts helps drain off static electricity charges before an electric arc takes place (flashover potential). Static grounding is often used in areas where the discharge (arcing) of the voltage buildup (static) can cause dangerous or undesirable conditions [500.4 FPN No. 3].

CAUTION: *Connecting metal parts to the earth (grounding) serves no purpose in electrical shock protection.*

(2) Equipment Bonding. Metal parts of electrical raceways, cables, enclosures, or equipment must be bonded together in a manner that creates a low-impedance path for ground-fault current to facilitate the operation of the circuit overcurrent device.

The fault current path must be capable of safely carrying the maximum ground-fault current likely to be imposed on it from any point on the wiring system where a ground fault may occur to the electrical supply source.

(3) Bonding Conductive Materials. Conductive materials such as metal water piping systems, metal sprinkler piping, metal gas piping, and other metal-piping systems, as well as exposed structural steel members likely to become energized must be bonded together in a manner that creates a low-impedance fault current path that is capable of carrying the maximum fault current likely to be imposed on it. **Figure 250–25**

Author's Comment: The phrase "likely to become energized" is subject to interpretation by the authority having jurisdiction.

(4) Fault Current Path. Electrical equipment, wiring, and other electrically conductive material likely to become energized must be installed in a manner that creates a low-impedance fault current path to facilitate the operation of overcurrent devices should a second ground fault from a different phase occur. **Figure 250–26**

Author's Comment: A single ground fault cannot be cleared on an ungrounded system because there's no low-impedance fault current path to the power source. The first ground fault simply grounds the previously ungrounded system. However, a second ground fault on a different phase results in a line-to-line short circuit between the two ground faults. The conductive path, between the ground faults, provides a low-impedance fault current path necessary so that the overcurrent device will open.

250.6 Objectionable Current.

(A) Preventing Objectionable Current. To prevent a fire, electric shock, or improper operation of circuit overcurrent devices or electronic equipment, electrical systems and equipment must be installed in a manner that prevents objectionable neutral current from flowing on metal parts.

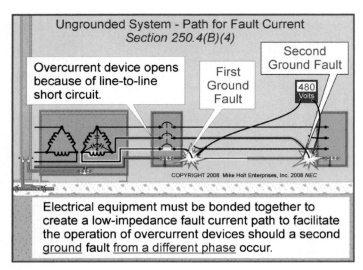

Figure 250–26

SPECIAL SECTION

Objectionable neutral current occurs because of improper neutral-to-case connections or wiring errors that violate 250.142(B).

Improper Neutral-to-Case Connection [250.142]

Panelboards. Objectionable neutral current will flow when the neutral conductor is connected to the metal case of a panelboard that is not used as service equipment. Figure 250–27

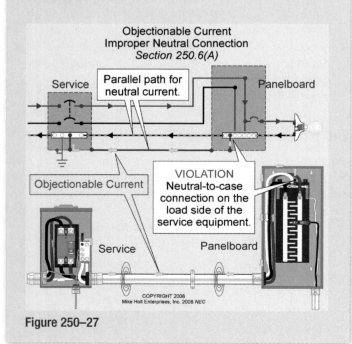

Figure 250–27

Separately Derived Systems. Objectionable neutral current will flow on conductive metal parts and conductors if the neutral conductor is connected to the circuit equipment grounding conductor on the load side of the system bonding jumper for separately derived system. Figures 250–28 and 250–29

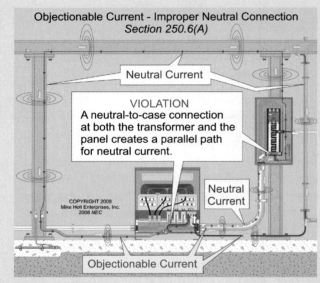

Figure 250–28

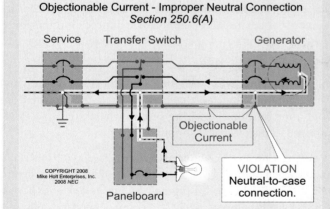

Figure 250–29

Disconnects. Objectionable neutral current will flow when the neutral conductor is connected to the metal case of a disconnecting means not part of service equipment. Figure 250–30

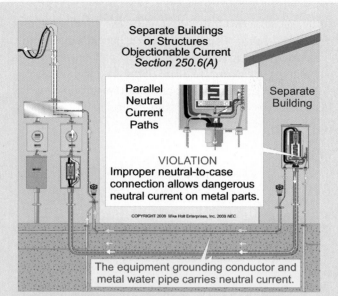

Figure 250–30

Wiring Errors. Objectionable neutral current will flow when the neutral conductor from one system is connected to a circuit of a different system. **Figure 250–31**

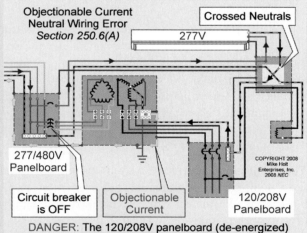

Figure 250–31

Objectionable neutral current will flow on metal parts when the circuit equipment grounding conductor is used as a neutral conductor:

- A 230V time-clock motor is replaced with a 115V time-clock motor, with the circuit equipment grounding conductor being used for neutral return current.

- A 115V water filter is wired to a 240V well-pump motor circuit, with the circuit equipment grounding conductor being used for neutral return current. **Figure 250–32**

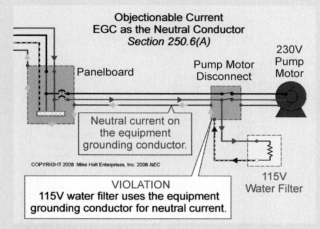

Figure 250–32

- Where the circuit equipment grounding conductor of Type NM cable is used for neutral return current. **Figure 250–33**

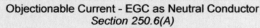

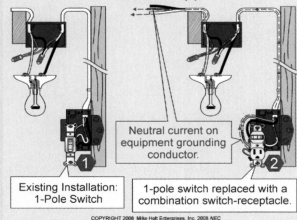

Figure 250–33

Dangers of Objectionable Current

Objectionable neutral current on metal parts can cause electric shock, fires, and improper operation of electronic equipment and overcurrent devices such as GFPs, GFCIs, and AFCIs..

Shock Hazard. When objectionable neutral current flows on metal parts, electric shock and even death can occur from elevated voltage on the metal parts. **Figure 250–34** and **250–35**

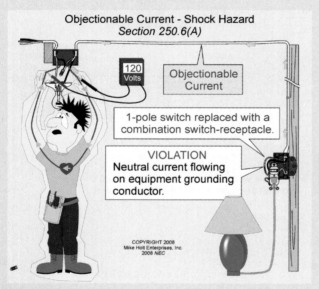

Figure 250–34

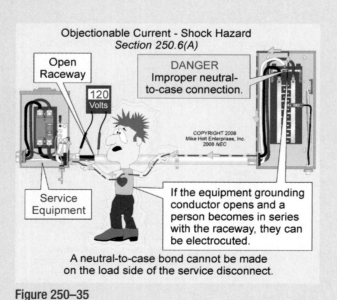

Figure 250–35

Fire Hazard. When objectionable neutral current flows on metal parts, a fire can ignite adjacent combustible material. Heat is generated whenever current flows, particularly over high-resistance parts. In addition, arcing at loose connections is especially dangerous in areas containing easily ignitable and explosive gases, vapors, or dust. **Figure 250–36**

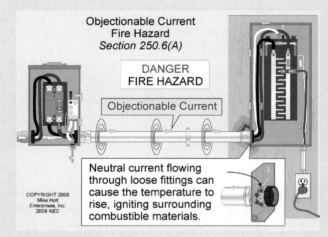

Figure 250–36

Improper Operation of Electronic Equipment. Objectionable neutral current flowing on metal parts of electrical equipment and building parts can cause electromagnetic fields which negatively affect the performance of electronic devices, particularly medical equipment. For more information, visit www.MikeHolt.com, click on the "Technical Link," then on "Power Quality." **Figure 250–37**

When objectionable neutral current travels on metal parts, a difference of potential will exist between all metal parts, which can cause some electronic equipment to operate improperly. **Figure 250–38** and **250–39**

Operation of Overcurrent Devices. When objectionable neutral current travels on metal parts, nuisance tripping of electronic overcurrent devices equipped with ground-fault protection can occur because some neutral current flows on the circuit equipment grounding conductor instead of the neutral conductor.

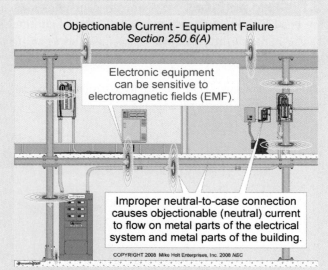

Objectionable Current - Equipment Failure
Section 250.6(A)

Electronic equipment can be sensitive to electromagnetic fields (EMF).

Improper neutral-to-case connection causes objectionable (neutral) current to flow on metal parts of the electrical system and metal parts of the building.

COPYRIGHT 2008 Mike Holt Enterprises, Inc. 2008 NEC

Figure 250–37

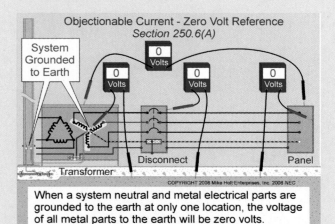

Objectionable Current - Zero Volt Reference
Section 250.6(A)

System Grounded to Earth

Disconnect Panel

Transformer

COPYRIGHT 2008 Mike Holt Enterprises, Inc. 2008 NEC

When a system neutral and metal electrical parts are grounded to the earth at only one location, the voltage of all metal parts to the earth will be zero volts.

Figure 250–38

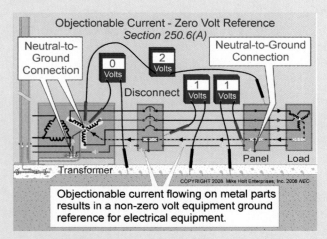

Objectionable Current - Zero Volt Reference
Section 250.6(A)

Neutral-to-Ground Connection

Neutral-to-Ground Connection

Disconnect

Panel Load

Transformer

COPYRIGHT 2008 Mike Holt Enterprises, Inc. 2008 NEC

Objectionable current flowing on metal parts results in a non-zero volt equipment ground reference for electrical equipment.

Figure 250–39

250.8 Termination of Grounding and Bonding Conductors.

(A) Permitted Methods. Grounding and bonding conductors must terminate in one of the following methods:

(1) Listed pressure connectors

(2) Terminal bars

(3) Pressure connectors listed for direct burial or concrete encasement [250.70]

(4) Exothermic welding

(5) Machine screws that engage at least two threads or are secured with a nut, **Figure 250–40**

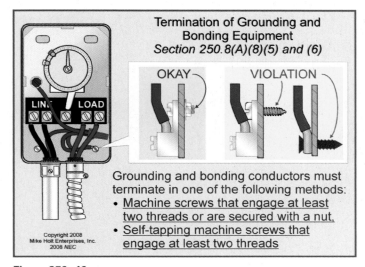

Termination of Grounding and Bonding Equipment
Section 250.8(A)(8)(5) and (6)

LINE   LOAD

OKAY   VIOLATION

Grounding and bonding conductors must terminate in one of the following methods:
• Machine screws that engage at least two threads or are secured with a nut,
• Self-tapping machine screws that engage at least two threads

Copyright 2008 Mike Holt Enterprises, Inc. 2008 NEC

Figure 250–40

(6) Self-tapping machine screws that engage at least two threads

(7) Connections that are part of a listed assembly

(8) Other listed means

250.10 Protection of Fittings. Grounding and bonding fittings must be protected from physical damage by:

(1) Locating the fittings so they aren't likely to be damaged.

(2) Enclosing the fittings in metal, wood, or equivalent protective covering.

> **Author's Comment:** Grounding and bonding fittings can be buried or encased in concrete if installed in accordance with 250.53(G), 250.68(A) Ex 1, and 250.70.

250.12 Clean Surfaces. Nonconductive coatings, such as paint, must be removed to ensure good electrical continuity, or the termination fittings must be designed so as to make such removal unnecessary [250.53(A) and 250.96(A)].

> **Author's Comment:** Tarnish on copper water pipe need not be removed before making a termination.

Practice Questions

PART I

Use the 2008 *NEC* to answer the following questions.

PART I. GENERAL PRACTICE QUESTIONS

1. The system bonding jumper is the connection between the _____ conductor and the equipment grounding conductor at a separately derived system.

   (a) grounded
   (b) ungrounded
   (c) grounding electrode
   (d) equipment bonding jumper

2. An effective ground-fault current path is an intentionally constructed, permanent, low-impedance path designed and intended to carry fault current from the point of a ground fault on a wiring system to _____.

   (a) ground
   (b) earth
   (c) the electrical supply source
   (d) none of these

3. A(n) _____ is an unintentional, electrically conducting connection between an ungrounded normally current-carrying conductor of an electrical circuit, and the normally non–current-carrying conductors, metallic enclosures, metallic raceways, metallic equipment, or earth.

   (a) grounded conductor
   (b) ground fault
   (c) equipment ground
   (d) bonding jumper

4. A ground-fault current path is an electrically conductive path from the point of a ground fault through normally non–current-carrying conductors, equipment, or the earth to the _____.

   (a) ground
   (b) earth
   (c) electrical supply source
   (d) none of these

5. Examples of ground-fault current paths include any combination of conductive materials including _____.

   (a) equipment grounding conductors
   (b) metallic raceways
   (c) metal water and gas piping
   (d) all of these

6. Grounded electrical systems shall be connected to earth in a manner that will _____.

   (a) limit voltages due to lightning, line surges, or unintentional contact with higher-voltage lines
   (b) stabilize the voltage-to-ground during normal operation
   (c) facilitate overcurrent device operation in case of ground faults
   (d) a and b

7. For grounded systems, normally non–current-carrying conductive materials enclosing electrical conductors or equipment shall be connected to earth so as to limit the voltage-to-ground on these materials.

   (a) True
   (b) False

8. For grounded systems, non–current-carrying conductive materials enclosing electrical conductors or equipment, or forming part of such equipment, shall be connected together and to the _____ to establish an effective ground-fault current path.

   (a) ground
   (b) earth
   (c) electrical supply source
   (d) none of these

9. In grounded systems, normally non–current-carrying electrically conductive materials that are likely to become energized shall be _____ in a manner that establishes an effective ground-fault current path.

    (a) connected together
    (b) connected to the electrical supply source
    (c) connected to the closest grounded conductor
    (d) a and b

10. For grounded systems, electrical equipment and conductive material likely to become energized, shall be installed in a manner that creates a _____ from any point on the wiring system where a ground fault may occur to the electrical supply source.

    (a) circuit facilitating the operation of the overcurrent device
    (b) low-impedance path
    (c) path capable of safely carrying the ground-fault current likely to be imposed on it
    (d) all of these

11. For grounded systems, electrical equipment and electrically conductive material likely to become energized, shall be installed in a manner that creates a low-impedance circuit capable of safely carrying the maximum ground-fault current likely to be imposed on it from where a ground fault may occur to the _____.

    (a) ground
    (b) earth
    (c) electrical supply source
    (d) none of these

12. For grounded systems, the earth is considered an effective ground-fault current path.

    (a) True
    (b) False

13. For ungrounded systems, non–current-carrying conductive materials enclosing electrical conductors or equipment shall be connected to the _____ in a manner that will limit the voltage imposed by lightning or unintentional contact with higher-voltage lines.

    (a) ground
    (b) earth
    (c) the electrical supply source
    (d) none of these

14. For ungrounded systems, non–current-carrying conductive materials enclosing electrical conductors or equipment, or forming part of such equipment, shall be connected together and to the supply system equipment in a manner that creates a low-impedance path for ground-fault current that is capable of carrying _____.

    (a) the maximum branch-circuit current
    (b) at least twice the maximum ground-fault current
    (c) the maximum fault current likely to be imposed on it
    (d) the equivalent to the main service rating

15. Electrically conductive materials that are likely to _____ in ungrounded systems shall be connected together and to the supply system grounded equipment in a manner that creates a low-impedance path for ground-fault current that is capable of carrying the maximum fault current likely to be imposed on it.

    (a) become energized
    (b) require service
    (c) be removed
    (d) be coated with paint or nonconductive materials

16. In ungrounded systems, electrical equipment, wiring, and other electrically conductive material likely to become energized shall be installed in a manner that creates a low-impedance circuit from any point on the wiring system to the electrical supply source to facilitate the operation of overcurrent devices should a(n) _____ fault from a different phase occur on the wiring system.

    (a) isolated ground
    (b) second ground
    (c) arc
    (d) high impedance

17. The grounding of electrical systems, circuit conductors, surge arresters, surge protective devices, and conductive normally non–current-carrying metal parts of equipment shall be installed and arranged in a manner that will prevent objectionable current.

    (a) True
    (b) False

18. Temporary currents resulting from accidental conditions, such as ground faults, are not considered to be objectionable currents.

    (a) True
    (b) False

19. Grounding conductors and bonding jumpers shall be connected by _____.

    (a) listed pressure connectors
    (b) terminal bars
    (c) exothermic welding
    (d) any of these

20. Ground clamps and fittings shall be protected from physical damage by being enclosed in _____ where there may be a possibility of physical damage.

    (a) metal
    (b) wood
    (c) the equivalent of a or b
    (d) none of these

21. _____ on equipment to be grounded shall be removed from contact surfaces to ensure good electrical continuity.

    (a) Paint
    (b) Lacquer
    (c) Enamel
    (d) any of these

250

Grounding and Bonding— Part II: System Grounding and Bonding

PART II. SYSTEM GROUNDING AND BONDING

250.20 Systems Required to be Grounded.

(A) Systems Below 50 Volts. Systems operating below 50V are not required to be grounded (connected to the earth) unless the transformers primary supply is from: **Figure 250–41**

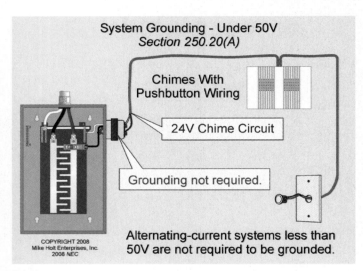

Figure 250–41

(1) A 277V or 480V system

(2) An ungrounded system

(B) Systems Over 50 Volts. The following systems must be connected (grounded) to the earth:

(1) Single-phase systems where the neutral conductor is used as a circuit conductor. **Figure 250–42**

(2) Three-phase, wye-connected systems where the neutral conductor is used as a circuit conductor. **Figure 250–43A**

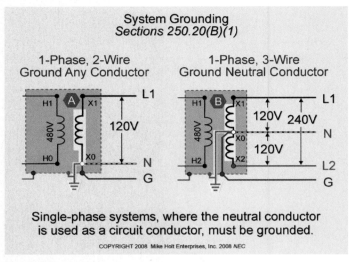

Figure 250–42

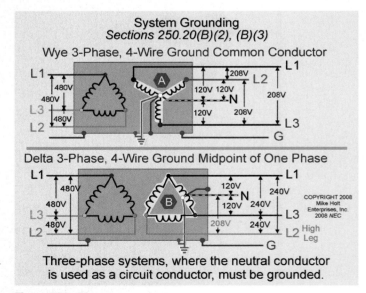

Figure 250–43

(3) Three-phase, high-leg delta-connected systems where the neutral conductor is used as a circuit conductor.

Figure 250–43B

(D) Separately Derived Systems. Separately derived systems as covered in 250.20(B) must be grounded (connected to the earth) in accordance with 250.30(A).

Author's Comment: A separately derived system is a wiring system whose power is derived from a source where there's no direct electrical connection to the supply conductors of another system. This includes most transformers because the primary circuit conductors don't have any direct electrical connection to the secondary circuit conductors [Article 100]. **Figure 250–44**

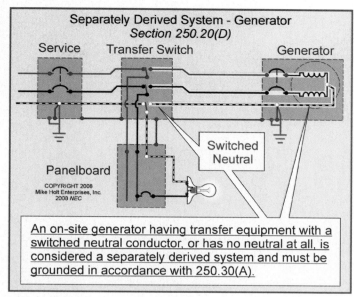

Figure 250–45

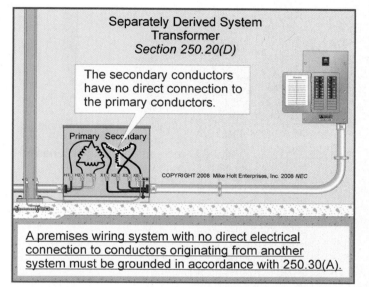

Figure 250–44

A generator having transfer equipment that switches the neutral conductor or has no neutral conductor at all must be grounded (connected to the earth) in accordance with 250.30(A). **Figure 250–45**

FPN No. 1: An alternate ac power source such as an on-site generator is not a separately derived system if the neutral conductor is solidly interconnected to a service-supplied system neutral conductor. An example would be a generator provided with a transfer switch that includes a neutral conductor that is not switched. **Figure 250–46**

250.24 Service Equipment—Grounding and Bonding.

(A) Grounded System. Service equipment supplied from a grounded system must have the neutral conductor terminate in accordance with (1) through (5).

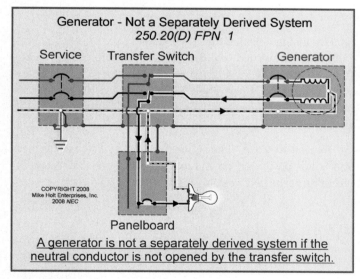

Figure 250–46

(1) Grounding Location. A grounding electrode conductor must connect the service neutral conductor to the grounding electrode at any accessible location, from the load end of the service drop or service lateral, up to and including the service disconnecting means. **Figure 250–47**

Author's Comment: Some inspectors require the service neutral conductor to be grounded (connected to the earth) from the meter socket enclosure, while other inspectors insist that the service neutral conductor be grounded (connected to the earth) only from the service disconnect.

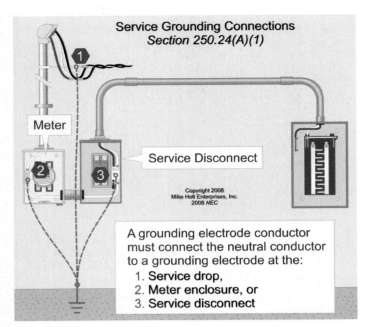

Service Grounding Connections
Section 250.24(A)(1)

Meter

Service Disconnect

Copyright 2008
Mike Holt Enterprises, Inc.
2008 NEC

A grounding electrode conductor must connect the neutral conductor to a grounding electrode at the:
1. Service drop,
2. Meter enclosure, or
3. Service disconnect

Figure 250–47

(4) Grounding Termination. When the service neutral conductor is connected to the service disconnecting means [250.24(B)] by a wire or busbar [250.28], the grounding electrode conductor is permitted to terminate to either the neutral terminal or the equipment grounding terminal within the service disconnect.

(5) Neutral-to-Case Connection. A neutral-to-case connection is not permitted on the load side of service equipment, except as permitted by 250.142(B). **Figure 250–48**

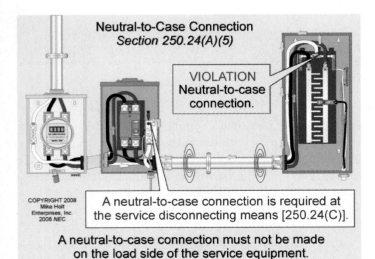

Neutral-to-Case Connection
Section 250.24(A)(5)

VIOLATION
Neutral-to-case connection.

COPYRIGHT 2008
Mike Holt
Enterprises, Inc.
2008 NEC

A neutral-to-case connection is required at the service disconnecting means [250.24(C)].

A neutral-to-case connection must not be made on the load side of the service equipment.

Figure 250–48

Author's Comment: If a neutral-to-case connection is made on the load side of service equipment, dangerous objectionable neutral current will flow on conductive metal parts of electrical equipment [250.6(A)]. Objectionable neutral current on metal parts of electrical equipment can cause electric shock and even death from ventricular fibrillation, as well as a fire. **Figures 250–49** and **250–50**

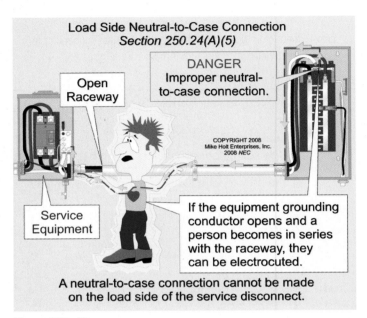

Load Side Neutral-to-Case Connection
Section 250.24(A)(5)

Open Raceway

DANGER
Improper neutral-to-case connection.

COPYRIGHT 2008
Mike Holt Enterprises, Inc.
2008 NEC

Service Equipment

If the equipment grounding conductor opens and a person becomes in series with the raceway, they can be electrocuted.

A neutral-to-case connection cannot be made on the load side of the service disconnect.

Figure 250–49

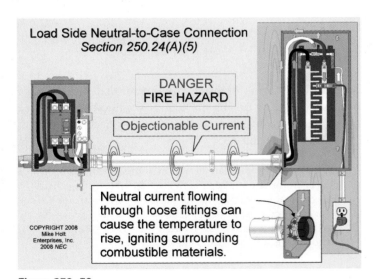

Load Side Neutral-to-Case Connection
Section 250.24(A)(5)

DANGER
FIRE HAZARD

Objectionable Current

COPYRIGHT 2008
Mike Holt
Enterprises, Inc.
2008 NEC

Neutral current flowing through loose fittings can cause the temperature to rise, igniting surrounding combustible materials.

Figure 250–50

(B) Bonding. A main bonding jumper [250.28] must be installed for the purpose of connecting the neutral conductor to the metal parts of the service disconnecting means. **Figure 250–51**

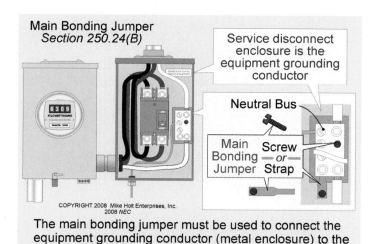

Figure 250–51

The main bonding jumper must be used to connect the equipment grounding conductor (metal enclosure) to the neutral conductor in a service disconnect.

Figure 250–51

(C) Neutral Conductor. A service neutral conductor from the electric utility must terminate to each service disconnecting means via a main bonding jumper [250.24(B)] that is installed between the service neutral conductor and each service disconnecting means enclosure. **Figures 250–52** and **250–53**

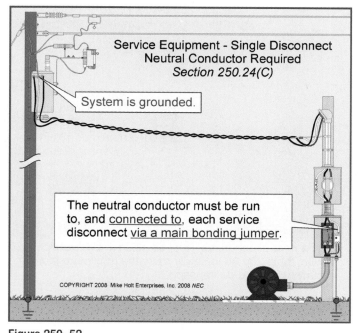

Figure 250–52

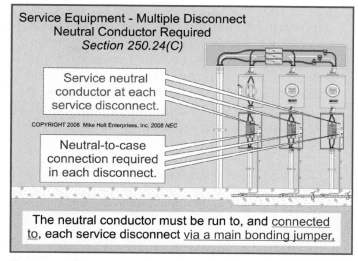

The neutral conductor must be run to, and connected to, each service disconnect via a main bonding jumper.

Figure 250–53

Author's Comment: The service neutral conductor provides the effective ground-fault current path to the power supply to ensure that dangerous voltage from a ground fault will be quickly removed by opening the overcurrent device [250.4(A)(3) and 250.4(A)(5)]. **Figures 250–54**

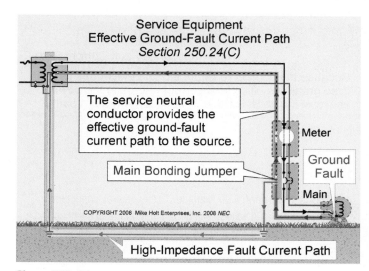

Figure 250–54

DANGER: *Dangerous voltage from a ground fault will not be removed from metal parts, metal piping, and structural steel if the service disconnecting means enclosure is not connected to the service neutral conductor. This is because the contact resistance of a grounding electrode to the earth is so great that insufficient fault current returns to the power supply if the earth is the only fault current return path to open the circuit overcurrent device.* **Figure 250–55**

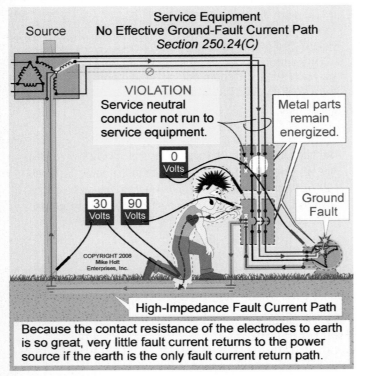

Service Equipment
No Effective Ground-Fault Current Path
Section 250.24(C)

VIOLATION
Service neutral conductor not run to service equipment.

Metal parts remain energized.

Ground Fault

High-Impedance Fault Current Path

Because the contact resistance of the electrodes to earth is so great, very little fault current returns to the power source if the earth is the only fault current return path.

Figure 250–55

Author's Comment: For example, if the neutral conductor is opened, dangerous voltage will be present on metal parts under normal conditions, providing the potential for electric shock. If the earth's ground resistance is 25 ohms and the load's resistance is 25 ohms, the voltage drop across each of these resistors will be half of the voltage source. Since the neutral is connected to the service disconnect, all metal parts will be elevated to 60V above the earth's potential for a 120/240V system. **Figure 250–56**

To determine the actual voltage on the metal parts from an open service neutral conductor, you need to do some complex math calculations. Visit www.MikeHolt.com and go to the "Free Stuff" link to download a spreadsheet for this purpose.

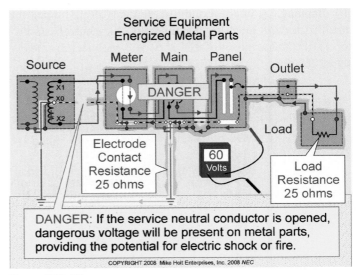

Service Equipment
Energized Metal Parts

Source | Meter | Main | Panel | Outlet

DANGER

Load

Electrode Contact Resistance 25 ohms

60 Volts

Load Resistance 25 ohms

DANGER: If the service neutral conductor is opened, dangerous voltage will be present on metal parts, providing the potential for electric shock or fire.

Figure 250–56

(1) Neutral Conductor Minimum Size. Because the service neutral conductor serves as the effective ground-fault current path to the source for ground faults, it must be sized so it can safely carry the maximum fault current likely to be imposed on it [110.10 and 250.4(A)(5)]. This is accomplished by sizing the neutral conductor in accordance with Table 250.66, based on the cross-sectional area of the ungrounded service conductor. **Figure 250–57**

Service Neutral Size
Section 250.24(C)(1)

4/0 AWG Service Phase Conductors

Minimum 2 AWG Service Neutral

Copyright 2008 Mike Holt Enterprises, Inc. 2008 NEC

The neutral conductor must be sized in accordance with Table 250.66 to safely carry the maximum fault current likely to be imposed.

Figure 250–57

Author's Comment: In addition, the neutral conductors must have the capacity to carry the maximum unbalanced neutral current in accordance with 220.61.

Question: What is the minimum size service neutral conductor required for a 480V, three-phase service where the ungrounded service conductors are sized at 500 kcmil and the maximum unbalanced load is 100A? **Figure 250–58**

(a) 3 AWG   (b) 2 AWG   (c) 1 AWG   (d) 1/0 AWG

Answer: (d) 1/0 AWG [Table 250.66]

The unbalanced load requires a 3 AWG service neutral conductor, which is rated 100A at 75ºC in accordance with Table 310.16 [220.61]. However, the service neutral conductor must not be smaller than 1/0 AWG to ensure it will accommodate the maximum fault current likely to be imposed on it [Table 250.66].

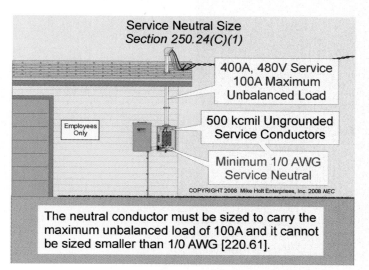

Figure 250–58

(2) Parallel Neutral Conductor. Where service conductors are paralleled, a neutral conductor must be installed in each of the parallel raceways and it must be sized in accordance with Table 250.66, based on the area of the largest service conductor in the raceway. In no case can the neutral conductor in each parallel set be sized smaller than 1/0 AWG [310.4(A)].

Author's Comment: In addition, the neutral conductors must have the capacity to carry the maximum unbalanced neutral current in accordance with 220.61

Question: What is the minimum size service neutral conductor required for a 480V, three-phase service installed in two raceways where the ungrounded service conductors in each of the raceways are 350 kcmil and the maximum unbalanced load is 100A? **Figure 250–59**

(a) 3 AWG   (b) 2 AWG   (c) 1 AWG   (d) 1/0 AWG

Answer: (d) 1/0 AWG per raceway [Table 250.66 and 310.4]

The unbalanced load of 50A in each raceways requires an 8 AWG service neutral conductor, which is rated 50A at 75ºC in accordance with Table 310.16 [220.61]. However, the smallest service neutral conductor permitted to be run in parallel in each raceway must not be smaller than 1/0 AWG [Table 250.66]

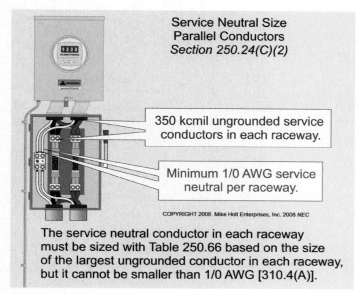

Figure 250–59

(D) Grounding Electrode Conductor. A grounding electrode conductor, sized in accordance with 250.66 based on the area of the ungrounded service conductor, must connect the metal parts of service equipment enclosures to a grounding electrode in accordance with Part III of Article 250.

Question: *What is the minimum size grounding electrode conductor for a 400A service where the ungrounded service conductors are sized at 500 kcmil?* **Figure 250–60**

(a) 3 AWG   (b) 2 AWG   (c) 1 AWG   (d) 1/0 AWG

Answer: *(d) 1/0 AWG [Table 250.66]*

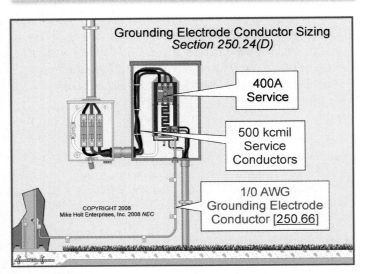

Figure 250–60

Author's Comment: Where the grounding electrode conductor is connected to a ground rod, that portion of the conductor that is the sole connection to the ground rod isn't required to be larger than 6 AWG copper [250.66(A)]. Where the grounding electrode conductor is connected to a concrete-encased electrode, that portion of the conductor that is the sole connection to the concrete-encased electrode isn't required to be larger than 4 AWG copper [250.66(B)]. **Figure 250–61** and **250–62**

250.28 Main Bonding Jumper and System Bonding Jumper. Main and system bonding jumpers must be installed as follows:

Author's Comments:

- **Main Bonding Jumper.** At service equipment, a main bonding jumper must be installed to electrically connect the neutral conductor to the service disconnect enclosure [250.24(B)]. **Figure 250–63**

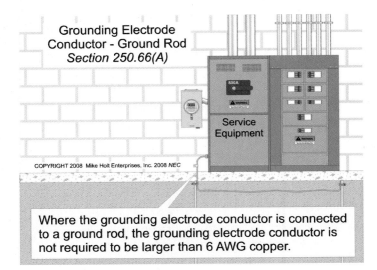

Where the grounding electrode conductor is connected to a ground rod, the grounding electrode conductor is not required to be larger than 6 AWG copper.

Figure 250–61

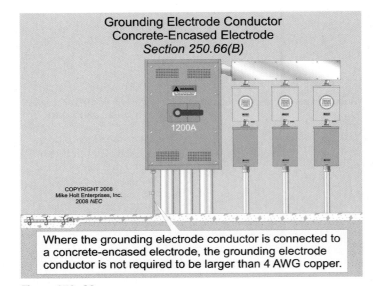

Where the grounding electrode conductor is connected to a concrete-encased electrode, the grounding electrode conductor is not required to be larger than 4 AWG copper.

Figure 250–62

The main bonding jumper provides the low-impedance path necessary for fault current to travel back to the power supply to open the circuit overcurrent device to clear a ground fault [250.24(C)]. **Figure 250–64**

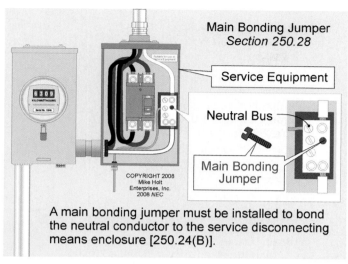

Main Bonding Jumper
Section 250.28

Service Equipment

Neutral Bus

Main Bonding Jumper

A main bonding jumper must be installed to bond the neutral conductor to the service disconnecting means enclosure [250.24(B)].

Figure 250–63

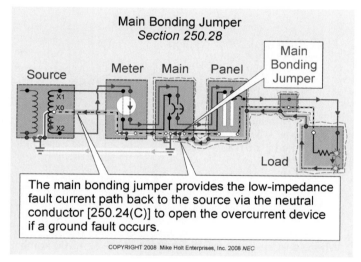

Main Bonding Jumper
Section 250.28

Main Bonding Jumper

Source   Meter   Main   Panel

Load

The main bonding jumper provides the low-impedance fault current path back to the source via the neutral conductor [250.24(C)] to open the overcurrent device if a ground fault occurs.

Figure 250–64

DANGER: *Metal parts of the electrical installation, as well as metal piping and structural steel, will become and remain energized with dangerous voltage from a ground fault if a main bonding jumper isn't installed at service equipment.* **Figure 250–65**

- **System Bonding Jumper.** A system bonding jumper is installed between the neutral terminal of a separately derived system and the circuit equipment grounding conductor of the secondary system [250.2 and 250.30(A)(1)]. **Figure 250–66**

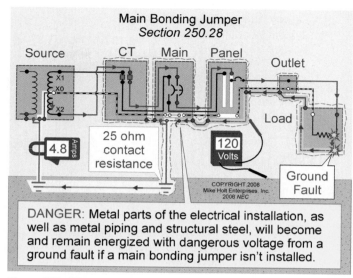

Main Bonding Jumper
Section 250.28

Source   CT   Main   Panel   Outlet

Load

4.8 Amps

25 ohm contact resistance

120 Volts

Ground Fault

DANGER: Metal parts of the electrical installation, as well as metal piping and structural steel, will become and remain energized with dangerous voltage from a ground fault if a main bonding jumper isn't installed.

Figure 250–65

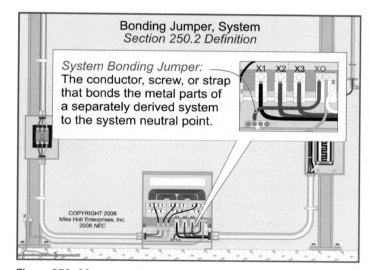

Bonding Jumper, System
Section 250.2 Definition

System Bonding Jumper: The conductor, screw, or strap that bonds the metal parts of a separately derived system to the system neutral point.

X1 X2 X3 X0

Figure 250–66

DANGER: *Metal parts of the electrical installation as well as metal piping and structural steel, will remain energized with dangerous voltage from a ground fault if a system bonding jumper isn't installed at a separately derived system.* **Figure 250–67**

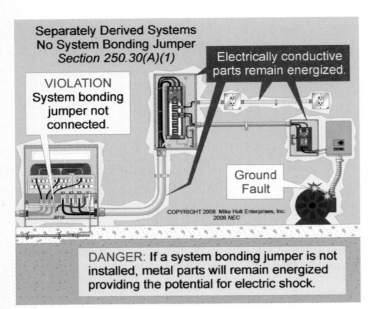

Figure 250–67

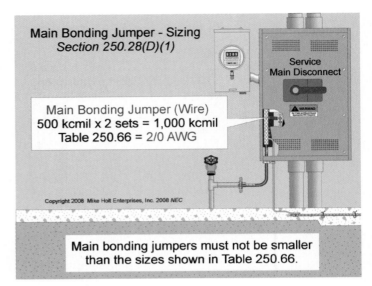

Figure 250–68

(A) Material. The bonding jumper can be a wire, bus, or screw.

(B) Construction. If the bonding jumper is a screw, it must be identified with a green finish visible with the screw installed.

(C) Attachment. Main and system bonding jumpers must terminate by one of the following means according to 250.8(A):

- Listed pressure connectors
- Terminal bars
- Pressure connectors listed as grounding and bonding equipment
- Exothermic welding
- Machine screw-type fasteners that engage not less than two threads or are secured with a nut
- Thread-forming machine screws that engage not less than two threads in the enclosure
- Connections that are part of a listed assembly
- Other listed means.

(D) Size. Main and system bonding jumpers must be sized not smaller than the sizes shown in Table 250.66. Where the service or secondary conductors have a total area larger than 1,100 kcmil copper or 1,750 kcmil aluminum, the bonding jumper must have an area not less than 12½ percent of the total conductor area of the largest ungrounded conductor. **Figures 250–68, 250–69,** and **250–70**

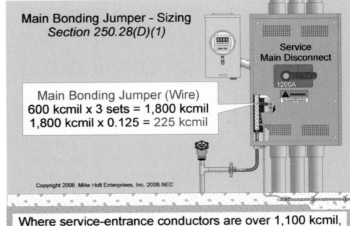

Figure 250–69

250.30 Separately Derived Systems—Grounding and Bonding.

Author's Comments:

- According to Article 100, a separately derived system is a wiring system whose power is derived from a source where there's no direct electrical connection to the supply conductors of another system.

- Transformers are considered separately derived when the primary conductors have no direct electrical connection to the secondary conductors. **Figure 250–71**

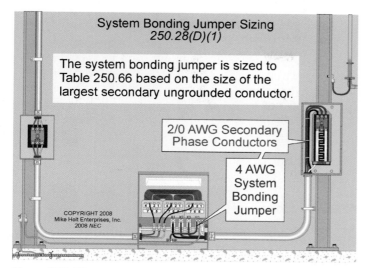

System Bonding Jumper Sizing
250.28(D)(1)

The system bonding jumper is sized to Table 250.66 based on the size of the largest secondary ungrounded conductor.

2/0 AWG Secondary Phase Conductors

4 AWG System Bonding Jumper

COPYRIGHT 2008 Mike Holt Enterprises, Inc. 2008 NEC

Figure 250–70

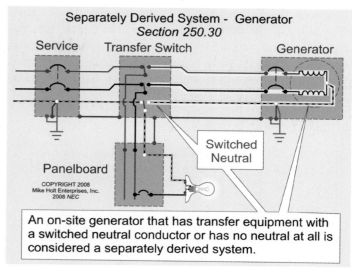

Separately Derived System - Generator
Section 250.30

Service   Transfer Switch   Generator

Switched Neutral

Panelboard

COPYRIGHT 2008 Mike Holt Enterprises, Inc. 2008 NEC

An on-site generator that has transfer equipment with a switched neutral conductor or has no neutral at all is considered a separately derived system.

Figure 250–72

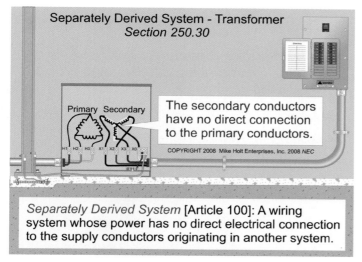

Separately Derived System - Transformer
Section 250.30

Primary Secondary

The secondary conductors have no direct connection to the primary conductors.

COPYRIGHT 2008 Mike Holt Enterprises, Inc. 2008 NEC

Separately Derived System [Article 100]: A wiring system whose power has no direct electrical connection to the supply conductors originating in another system.

Figure 250–71

- Generators that supply transfer equipment that switches the neutral conductor is an example of a separately derived system. **Figure 250–72**

(A) Grounded Systems. Separately derived systems must be grounded (connected to the earth) and bonded in accordance with (1) through (8). A neutral-to-case connection must not be on the load side of the system bonding jumper, except as permitted by 250.142(B).

CAUTION: *Dangerous objectionable neutral current will flow on conductive metal parts of electrical equipment as well as metal piping and structural steel, in violation of 250.6(A), if more than one system bonding jumper is installed, or if it's not located where the grounding electrode conductor terminates to the neutral conductor.* **Figure 250–73**

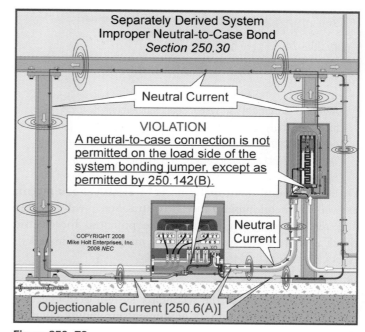

Separately Derived System
Improper Neutral-to-Case Bond
Section 250.30

Neutral Current

VIOLATION
A neutral-to-case connection is not permitted on the load side of the system bonding jumper, except as permitted by 250.142(B).

Neutral Current

COPYRIGHT 2008 Mike Holt Enterprises, Inc. 2008 NEC

Objectionable Current [250.6(A)]

Figure 250–73

(1) System Bonding Jumper. A system bonding jumper must be installed at the same location where the grounding electrode conductor terminates to the neutral terminal of the separately derived system; either at the separately derived system or the system disconnecting means, but not at both locations [250.30(A)(3)]. **Figure 250–74**

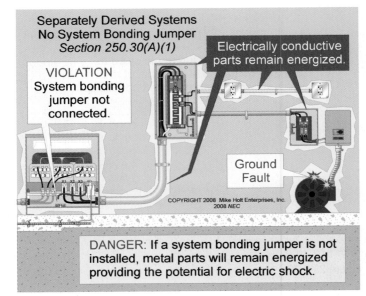

Separately Derived Systems
No System Bonding Jumper
Section 250.30(A)(1)

VIOLATION
System bonding jumper not connected.

Electrically conductive parts remain energized.

Ground Fault

DANGER: If a system bonding jumper is not installed, metal parts will remain energized providing the potential for electric shock.

Figure 250–75

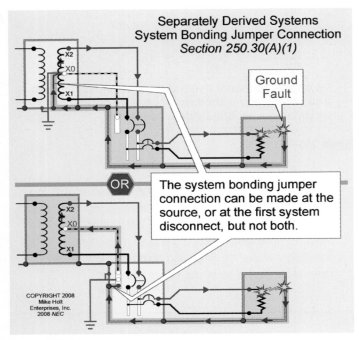

Separately Derived Systems
System Bonding Jumper Connection
Section 250.30(A)(1)

Ground Fault

OR

The system bonding jumper connection can be made at the source, or at the first system disconnect, but not both.

Figure 250–74

Question: What size equipment bonding jumper is required for flexible metal conduit containing 300 kcmil secondary conductors? **Figure 250–76**

(a) 3 AWG   (b) 2 AWG   (c) 1 AWG   (d) 1/0 AWG

Answer: (b) 2 AWG [Table 250.66]

Author's Comment: A system bonding jumper is a conductor, screw, or strap that bonds the metal parts of a separately derived system to a system neutral point [250.2] and it's sized to Table 250.66 in accordance with 250.28(D).

During a ground fault, metal parts of electrical equipment, as well as metal piping and structural steel, will become and remain energized providing the potential for electric shock and fire if the system bonding jumper is not installed. **Figure 250–75**

(2) Equipment Bonding Jumper Size. An equipment bonding jumper must be run to the secondary system disconnecting means. Where the secondary equipment grounding conductor is of the wire type, it must be sized in accordance with Table 250.66, based on the area of the largest ungrounded secondary conductor in the raceway or cable.

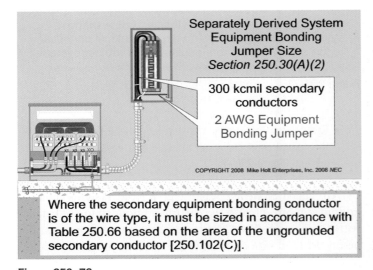

Separately Derived System
Equipment Bonding
Jumper Size
Section 250.30(A)(2)

300 kcmil secondary conductors

2 AWG Equipment Bonding Jumper

Where the secondary equipment bonding conductor is of the wire type, it must be sized in accordance with Table 250.66 based on the area of the ungrounded secondary conductor [250.102(C)].

Figure 250–76

(3) Grounding—Single Separately Derived System. A grounding electrode conductor must connect the neutral terminal of a separately derived system to a grounding electrode of a type identified in 250.30(A)(7). The grounding electrode conductor must be sized in accordance with 250.66, based on the area of the ungrounded secondary conductor. **Figure 250–77**

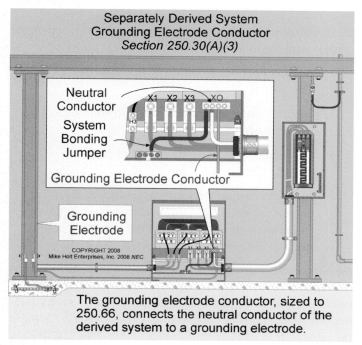

The grounding electrode conductor, sized to 250.66, connects the neutral conductor of the derived system to a grounding electrode.

Figure 250–77

Author's Comments:

- System grounding is intended to reduce overvoltage caused by induction from indirect lightning, or restriking/intermittent ground faults. Induced voltage imposed from lightning can be reduced by short grounding conductors and eliminating unnecessary bends and loops [250.4(A)(1) FPN]. **Figure 250–78**

- System grounding also helps reduce fires in buildings as well as voltage stress on electrical insulation, thereby ensuring longer insulation life for motors, transformers, and other system components.

To prevent objectionable neutral current from flowing [250.6] onto metal parts, the grounding electrode conductor must originate at the same point on the separately derived system where the system bonding jumper is connected [250.30(A)(1)].

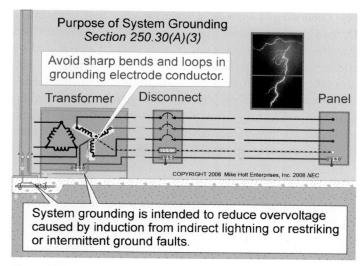

System grounding is intended to reduce overvoltage caused by induction from indirect lightning or restriking or intermittent ground faults.

Figure 250–78

Exception No. 1: Where the system bonding jumper is a wire or busbar, the grounding electrode conductor is permitted to terminate to either the neutral terminal or the equipment grounding terminal, bar, or bus in accordance with 250.30(A)(1). **Figure 250–79**

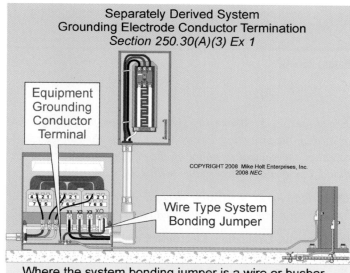

Where the system bonding jumper is a wire or busbar, the grounding electrode conductor can terminate to the equipment grounding terminal of the derived system.

Figure 250–79

Exception No. 3: Separately derived systems rated 1 kVA or less are not required to be grounded (connected to the earth).

(4) Grounding Electrode Conductor, Multiple Separately Derived Systems. Where there are multiple separately derived systems, the neutral terminal of each derived system can be connected to a common grounding electrode conductor. The grounding electrode conductor and grounding electrode conductor tap must comply with (a) through (c). The grounding electrode conductor and its taps must terminate at the same point on the separately derived system where the system bonding jumper is connected. **Figure 250–80**

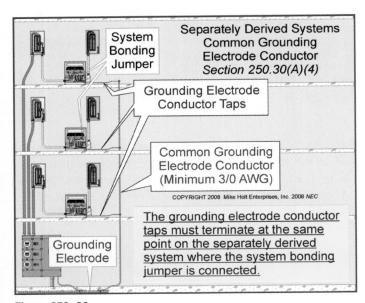

Figure 250–80

Exception No. 1: Where the system bonding jumper is a wire or busbar, the grounding electrode conductor tap can terminate to either the neutral terminal or the equipment grounding terminal, bar, or bus in accordance with 250.30(A)(1).

Exception No. 2: Separately derived systems rated 1 kVA or less are not required to be grounded (connected to the earth).

(a) Common Grounding Electrode Conductor. The common grounding electrode conductor must not be smaller than 3/0 AWG copper.

(b) Tap Conductor Size. Grounding electrode conductor taps must be sized in accordance with Table 250.66, based on the area of the largest ungrounded secondary conductor.

(c) Connections. Grounding electrode conductor tap connections must be made at an accessible location by:

(1) Listed connector.

(2) Listed connections to aluminum or copper busbars not less than ¼ x 2 in. Where aluminum busbars are used, the installation must comply with 250.64(A).

(3) Exothermic welding.

Grounding electrode conductor taps must be connected to the common grounding electrode conductor so the common grounding electrode conductor isn't spliced.

(5) Installation. The grounding electrode conductor must comply with the following:

- Be of copper where within 18 in. of earth [250.64(A)].

- Securely fastened to the surface on which it's carried [250.64(B)].

- Adequately protected if exposed to physical damage [250.64(B)].

- Metal enclosures enclosing a grounding electrode conductor must be made electrically continuous from the point of attachment to cabinets or equipment to the grounding electrode [250.64(E)].

(6) Structural Steel and Metal Piping. To ensure dangerous voltage from a ground fault is removed quickly, structural steel and metal piping in the area served by a separately derived system must be connected to the neutral conductor at the separately derived system in accordance with 250.104(D).

(7) Grounding Electrode. The grounding electrode must be as close as possible, and preferably in the same area where the system bonding jumper is located and be one of the following: **Figure 250–81**

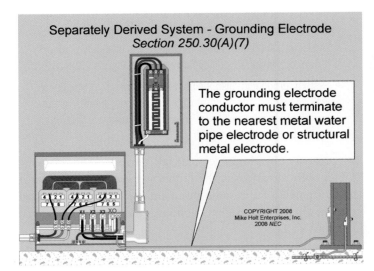

Figure 250–81

(1) Metal water pipe electrode, within 5 ft of entry to the building [250.52(A)(1)].

(2) Metal building frame electrode [250.52(A)(2)].

Exception No. 1: Where the electrodes listed in (1) or (2) are not available, one of the following must be used:

- *A concrete-encased electrode encased by not less than 2 in. of concrete, located horizontally near the bottom or vertically, and within that portion of concrete foundation or footing that is in direct contact with the earth, consisting of at least 20 ft of electrically conductive steel reinforcing bars or rods of not less than ½ in. diameter [250.52(A)(3)].*

- *A ground ring electrode encircling the building or structure, buried not less than 30 in. below grade, consisting of at least 20 ft of bare copper conductor not smaller than 2 AWG [250.52(A)(4) and 250.53(F)].*

- *A ground rod electrode having not less than 8 ft of contact with the soil meeting the requirements of 250.56 [250.52(A)(5) and 250.53(G)].*

- *Other metal underground systems, piping systems, or underground tanks [250.52(A)(8)].*

(8) System Bonding at Disconnect. Where the system bonding jumper is installed at the disconnecting means instead of at the source, the following requirements apply: **Figure 250–82**

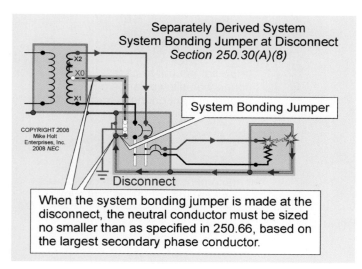

Separately Derived System
System Bonding Jumper at Disconnect
Section 250.30(A)(8)

System Bonding Jumper

COPYRIGHT 2008
Mike Holt
Enterprises, Inc.
2008 NEC

Disconnect

When the system bonding jumper is made at the disconnect, the neutral conductor must be sized no smaller than as specified in 250.66, based on the largest secondary phase conductor.

Figure 250–82

(a) Routing and Sizing. Because the secondary neutral conductor serves as the effective ground-fault current path for ground-fault current, it must be routed with the secondary conductors and sized not smaller than specified in Table 250.66, based on the area of the secondary conductor.

(b) Parallel Conductors. If the secondary conductors are installed in parallel, the secondary neutral conductor in each raceway or cable must be sized not smaller than specified in Table 250.66, based on the area of the largest ungrounded conductor in the raceway or cable. In no case is the neutral conductor permitted to be smaller than 1/0 AWG [310.4].

> **Author's Comment:** Where the system bonding jumper is installed at the disconnecting means instead of at the source, an equipment bonding conductor must connect the metal parts of the separately derived system to the neutral conductor at the disconnecting means in accordance with 250.30(A)(2).

250.32 Buildings or Structures Supplied by a Feeder or Branch Circuit.

(A) Grounding Electrode. Each building or structure's disconnect must be connected to an electrode of a type identified in 250.52. **Figure 250–83**

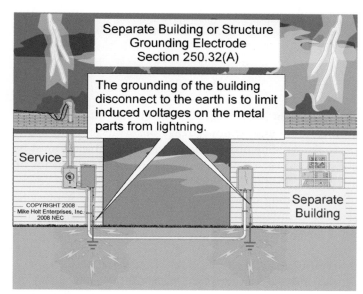

Separate Building or Structure
Grounding Electrode
Section 250.32(A)

The grounding of the building disconnect to the earth is to limit induced voltages on the metal parts from lightning.

Service

COPYRIGHT 2008
Mike Holt Enterprises, Inc.
2008 NEC

Separate
Building

Figure 250–83

Author's Comments:

• The grounding of the building or structure disconnecting means to the earth is intended to help in limiting induced voltages on the metal parts from nearby lightning strikes [250.4(A)(1)].

• The *Code* prohibits the use of the earth to serve as an effective ground-fault current path [250.4(A)(5) and 250.4(B)(4)].

Exception: A grounding electrode isn't required where the building or structure is served with a 2-wire, 3-wire, or 4-wire branch circuit. **Figure 250–84**

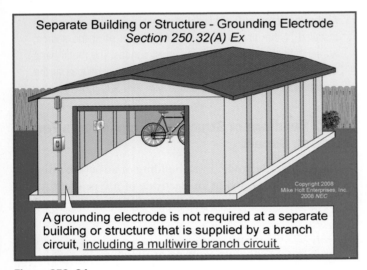

Separate Building or Structure - Grounding Electrode
Section 250.32(A) Ex

A grounding electrode is not required at a separate building or structure that is supplied by a branch circuit, including a multiwire branch circuit.

Figure 250–84

(B) Equipment Grounding Conductor. To quickly clear a ground fault and remove dangerous voltage from metal parts, the building or structure disconnecting means must be connected to the circuit equipment grounding conductor of a type described in 250.118. Where the supply circuit equipment grounding conductor is of the wire type, it must be sized in accordance with 250.122, based on the rating of the supply overcurrent device rating. **Figure 250–85**

CAUTION: *To prevent dangerous objectionable neutral current from flowing onto metal parts [250.6(A)], the supply circuit neutral conductor is not permitted to be connected to the remote building or structure disconnecting means [250.142(B)].* **Figure 250–86**

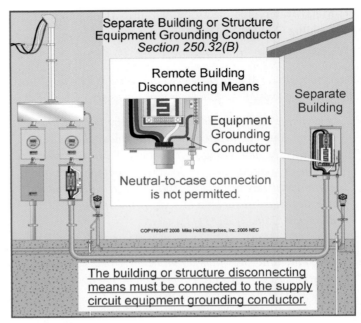

Separate Building or Structure
Equipment Grounding Conductor
Section 250.32(B)

Remote Building Disconnecting Means

Separate Building

Equipment Grounding Conductor

Neutral-to-case connection is not permitted.

COPYRIGHT 2008 Mike Holt Enterprises, Inc. 2008 NEC

The building or structure disconnecting means must be connected to the supply circuit equipment grounding conductor.

Figure 250–85

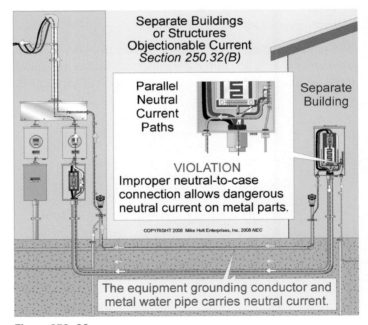

Separate Buildings or Structures
Objectionable Current
Section 250.32(B)

Parallel Neutral Current Paths

Separate Building

VIOLATION
Improper neutral-to-case connection allows dangerous neutral current on metal parts.

COPYRIGHT 2008 Mike Holt Enterprises, Inc. 2008 NEC

The equipment grounding conductor and metal water pipe carries neutral current.

Figure 250–86

Exception: For existing premises, when an equipment grounding conductor was not run to the building or structure disconnecting means, the building or structure disconnecting means can remain connected to the neutral conductor where there are no continuous metallic paths between buildings and structures, ground-fault protection of equipment isn't installed on the supply side of the circuit, and the neutral conductor is sized no smaller than the larger of:

(1) The maximum unbalanced neutral load in accordance with 220.61.

(2) The requirements of 250.122, based on the rating of the circuit overcurrent device

(E) Grounding Electrode Conductor. The grounding electrode conductor must terminate to the grounding terminal of the disconnecting means, and it must be sized in accordance with 250.66, based on the conductor area of the ungrounded feeder conductor.

Question: *What size grounding electrode conductor is required for a building disconnect supplied with a 3/0 AWG feeder?* **Figure 250–87**

(a) 4 AWG   (b) 3 AWG   (c) 2 AWG   (d) 1 AWG

Answer: *(a) 4 AWG [Table 250.66]*

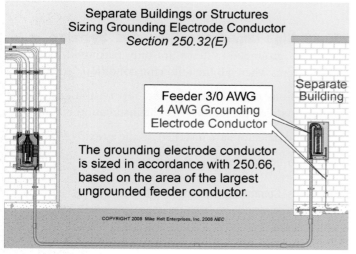

Figure 250–87

Author's Comment: Where the grounding electrode conductor is connected to a ground rod, that portion of the conductor that is the sole connection to the ground rod isn't required to be larger than 6 AWG copper [250.66(A)]. Where the grounding electrode conductor is connected to a concrete-encased electrode, that portion of the conductor that is the sole connection to the concrete-encased electrode isn't required to be larger than 4 AWG copper [250.66(B)].

250.34 Generators—Portable and Vehicle-Mounted.

(A) Portable Generators. The frame of a portable generator isn't required to be grounded (connected to the earth) if: **Figure 250–88**

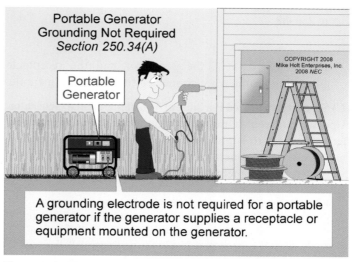

Figure 250–88

(1) The generator only supplies equipment or receptacles mounted on the generator, and

(2) The metal parts of the generator and the receptacle grounding terminal are connected to the generator frame.

(B) Vehicle-Mounted Generators. The frame of a vehicle-mounted generator isn't required to be grounded (connected to the earth) if: **Figure 250–89**

(1) The generator frame is bonded to the vehicle frame,

(2) The generator only supplies equipment or receptacles mounted on the vehicle or generator, and

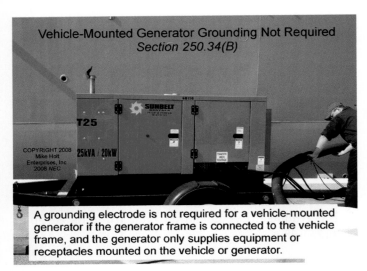

Vehicle-Mounted Generator Grounding Not Required
Section 250.34(B)

A grounding electrode is not required for a vehicle-mounted generator if the generator frame is connected to the vehicle frame, and the generator only supplies equipment or receptacles mounted on the vehicle or generator.

Figure 250–89

(3) The metal parts of the generator and the receptacle grounding terminal are connected to the generator frame.

(C) Separately Derived Portable or Vehicle-Mounted Generator. A portable or vehicle-mounted generator used as a separately derived system to supply equipment or receptacles mounted on the vehicle or generator must have the neutral conductor connected to the generator frame.

> **FPN:** A portable or vehicle-mounted generator supplying fixed wiring of premises must be grounded (connected to the earth) and bonded in accordance with 250.30 for separately derived systems and 250.35 for nonseparately derived systems.

250.35 Permanently Installed Generators.

(A) Separately Derived System. Where the generator is installed as a separately derived system, the system must be grounded (connected to the earth) and bonded in accordance with 250.30.

(B) Nonseparately Derived System. An equipment bonding jumper must be installed according to (1) or (2) as follows.

(1) Supply Side of Disconnect. An equipment bonding jumper must be installed from the generator equipment grounding terminal to the equipment grounding terminal of the generator disconnecting means. The bonding jumper must be of the wire type and be sized in accordance with Table 250.66, based on the cross-sectional area of the ungrounded generator conductors [250.102(C)].

(2) Load Side. The circuit equipment grounding conductor on the load side of each generator overcurrent device is required [250.86], and where of the wire type, it must be sized in accordance with 250.122, based on the rating of the overcurrent device [250.102(D)]. **Figure 250–90**

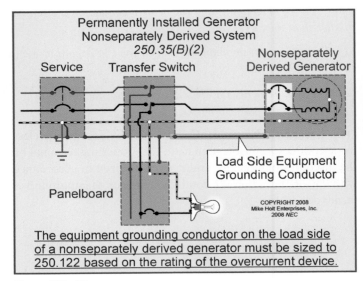

Permanently Installed Generator
Nonseparately Derived System
250.35(B)(2)

Service   Transfer Switch   Nonseparately Derived Generator

Load Side Equipment Grounding Conductor

Panelboard

COPYRIGHT 2008
Mike Holt Enterprises, Inc.
2008 NEC

The equipment grounding conductor on the load side of a nonseparately derived generator must be sized to 250.122 based on the rating of the overcurrent device.

Figure 250–90

Author's Comment: The frame of a nonseparately derived system generator is not required to be connected to a grounding electrode.

250.36 High-Impedance Grounded Systems. High-impedance grounded systems are only permitted for three-phase systems where all the following conditions are met:

(1) Conditions of maintenance and supervision ensure that only qualified persons service the installation.

(2) Ground detectors are installed on the system [250.21(B)].

(3) Line-to-neutral loads aren't served.

Author's Comment: High-impedance grounded systems are generally referred to as "high-resistance grounded systems" in the industry.

(A) Grounding Impedance Location. To limit fault current to a very low value, high-impedance grounded systems have a resistor installed between the neutral point of the derived system and the grounding electrode conductor. **Figure 250–91**

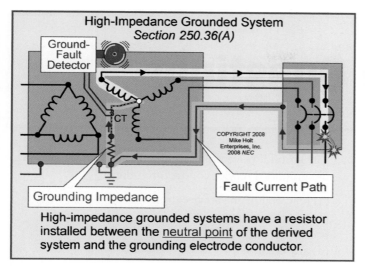

High-Impedance Grounded System
Section 250.36(A)

Ground-Fault Detector

CT

COPYRIGHT 2008
Mike Holt
Enterprises, Inc.
2008 NEC

Fault Current Path

Grounding Impedance

High-impedance grounded systems have a resistor installed between the <u>neutral point</u> of the derived system and the grounding electrode conductor.

Figure 250–91

FPN: For more information on this topic see IEEE 142, *Recommended Practice for Grounding of Industrial and Commercial Power Systems* (Green Book).

PART II Practice Questions

Use the 2008 *NEC* to answer the following questions.

PART II. SYSTEM GROUNDING PRACTICE QUESTIONS

1. AC circuits of less than 50V shall be grounded if supplied by a transformer whose supply system exceeds 150 volts-to-ground.

    (a) True
    (b) False

2. AC systems of 50V to 1,000V that supply premises wiring systems shall be grounded where the system is three-phase, 4-wire, wye-connected with the neutral conductor used as a circuit conductor.

    (a) True
    (b) False

3. AC systems of 50V to 1,000V that supply premises wiring systems shall be grounded where supplied by a three-phase, 4-wire, delta-connected system in which the midpoint of one phase winding is used as a circuit conductor.

    (a) True
    (b) False

4. An alternate ac power source such as an on-site generator is not a separately derived system if the _____ is solidly interconnected to a service-supplied system grounded conductor.

    (a) ignition system
    (b) fuel cell
    (c) grounded conductor
    (d) line conductor

5. _____ AC systems operating at 480V shall have ground detectors installed on the system.

    (a) Grounded
    (b) Solidly grounded
    (c) Effectively grounded
    (d) Ungrounded

6. The grounding electrode conductor shall be connected to the grounded service conductor at the _____.

    (a) load end of the service drop
    (b) meter equipment
    (c) service disconnect
    (d) any of these

7. Where the main bonding jumper is installed from the grounded conductor terminal bar to the equipment grounding terminal bar in service equipment, the _____ conductor is permitted to be connected to the equipment grounding terminal bar.

    (a) grounding
    (b) grounded
    (c) grounding electrode
    (d) none of these

8. For a grounded system, an unspliced _____ shall be used to connect the equipment grounding conductor(s) and the service disconnecting means to the grounded conductor of the system within the enclosure for each service disconnect.

    (a) grounding electrode
    (b) main bonding jumper
    (c) busbar
    (d) insulated copper conductor

9. Where an ac system operating at less than 1,000V is grounded at any point, the _____ conductor(s) shall be run to each service disconnecting means and shall be connected to each disconnecting means grounded conductor(s) terminal or bus.

    (a) ungrounded
    (b) grounded
    (c) grounding
    (d) none of these

10. The grounded conductor brought to service equipment shall be routed with the phase conductors and shall not be smaller than specified in Table _____ when the service-entrance conductors are not larger than 1,100 kcmil copper.

(a) 250.66
(b) 250.122
(c) 310.16
(d) 430.52

11. When service-entrance conductors exceed 1,100 kcmil for copper, the required grounded conductor for the service shall be sized not less than _____ percent of the area of the largest ungrounded service-entrance conductor.

(a) 9
(b) 11
(c) 12½
(d) 15

12. Where service-entrance phase conductors are installed in parallel, the size of the grounded conductor in each raceway shall be based on the size of the ungrounded service-entrance conductor in the raceway, but not smaller than _____.

(a) 1/0 AWG
(b) 2/0 AWG
(c) 3/0 AWG
(d) 4/0 AWG

13. A grounding electrode conductor, sized in accordance with 250.66, shall be used to connect the equipment grounding conductors, the service-equipment enclosures, and, where the system is grounded, the grounded service conductor to a grounding electrode.

(a) True
(b) False

14. A main bonding jumper shall be a _____ or similar suitable conductor.

(a) wire
(b) bus
(c) screw
(d) any of these

15. Where a main bonding jumper is a screw only, the screw shall be identified by a(n) _____ that shall be visible with the screw installed.

(a) silver or white finish
(b) etched ground symbol
(c) hexagonal head
(d) green finish

16. Main bonding jumpers and system bonding jumpers shall not be smaller than the sizes shown in _____.

(a) Table 250.66
(b) Table 250.122
(c) Table 310.16
(d) Chapter 9, Table 8

17. Where the supply conductors are larger than 1,100 kcmil copper or 1,750 kcmil aluminum, the main bonding jumper shall have an area that is _____ the area of the largest phase conductor when of the same material.

(a) at least equal to
(b) at least 50 percent of
(c) not less than 12½ percent of
(d) not more than 12½ percent of

18. A grounded conductor shall not be connected to normally non–current-carrying metal parts of equipment on the _____ side of the point of grounding of the separately derived system except as otherwise permitted in Article 250.

(a) supply
(b) grounded
(c) high voltage
(d) load

19. An unspliced _____ that is sized based on the derived phase conductors shall be used to connect the equipment grounding conductors of a separately derived system to the grounded conductor.

(a) system bonding jumper
(b) equipment grounding conductor
(c) grounded conductor
(d) grounding electrode conductor

20. The connection of the system bonding jumper for a separately derived system shall be made _____ on the separately derived system from the source to the first system disconnecting means or overcurrent device.

    (a) in at least two locations
    (b) in every location that the grounded conductor is present
    (c) at any single point
    (d) none of these

21. Where an equipment bonding jumper of the wire type is run with the derived phase conductors from the source of a separately derived system to the first disconnecting means, it shall be sized in accordance with 250.102(C), based on _____.

    (a) the size of the primary conductors
    (b) the size of the secondary overcurrent protection
    (c) the size of the derived phase conductors
    (d) one third the size of the primary grounded conductor

22. For a single separately derived system, the grounding electrode conductor connects the grounding electrode to the grounded conductor of the derived system at the same point on the separately derived system where the _____ is connected.

    (a) metering equipment
    (b) transfer switch
    (c) system bonding jumper
    (d) largest circuit breaker

23. The grounding electrode conductor for a single separately derived system is used to connect the grounded conductor of the derived system to the grounding electrode.

    (a) True
    (b) False

24. Grounding electrode conductor taps from a separately derived system to a common grounding electrode conductor are permitted when a building or structure has multiple separately derived systems, provided that the taps terminate at the same point as the system bonding jumper.

    (a) True
    (b) False

25. The common grounding electrode conductor installed for multiple separately derived systems shall not be smaller than _____.

    (a) 1/0 AWG
    (b) 2/0 AWG
    (c) 3/0 AWG
    (d) 4/0 AWG

26. Each tap conductor to a common grounding electrode conductor for multiple separately derived systems shall be sized in accordance with _____ based on the derived phase conductors of the separately derived system it serves.

    (a) 250.66
    (b) 250.118
    (c) 250.122
    (d) 310.15

27. Tap connections to the common grounding electrode conductor for multiple separately derived systems shall be made at an accessible location by _____.

    (a) a listed connector
    (b) listed connections to aluminum or copper busbars
    (c) by the exothermic welding process
    (d) any of these

28. Tap connections to a common grounding electrode conductor for multiple separately derived systems may be made to a copper or aluminum busbar that is _____.

    (a) not over ½ in. x 4 in.
    (b) not over ¼ in. x 2 in.
    (c) at least ¼ in. x 2 in.
    (d) a and c

29. In an area served by a separately derived system, the _____ shall be connected to the grounded conductor of the separately derived system.

    (a) structural steel
    (b) metal piping
    (c) metal building skin
    (d) a and b

30. The grounding electrode for a separately derived system shall be as near as practicable to, and preferably in the same area as, the grounding electrode conductor connection to the system.

    (a) True
    (b) False

31. A grounding electrode shall be required if a building or structure is supplied by a feeder.

    (a) True
    (b) False

32. A grounding electrode at a separate building or structure shall be required where one multiwire branch circuit serves the building or structure.

    (a) True
    (b) False

33. When supplying a grounded system at a separate building or structure, an equipment grounding conductor shall be run with the supply conductors and connected to the building disconnecting means.

    (a) True
    (b) False

34. The size of the grounding electrode conductor for a building or structure supplied by a feeder shall not be smaller than that identified in _____, based on the largest ungrounded supply conductor.

    (a) 250.66
    (b) 250.122
    (c) Table 310.16
    (d) none of these

35. The frame of a portable generator shall not be required to be connected to a(n) _____ if the generator only supplies equipment mounted on the generator or cord-and-plug connected equipment using receptacles mounted on the generator.

    (a) grounding electrode
    (b) grounded conductor
    (c) ungrounded conductor
    (d) equipment grounding conductor

36. The frame of a vehicle-mounted generator shall not be required to be connected to a(an) _____ if the generator only supplies equipment mounted on the vehicle or cord-and-plug connected equipment using receptacles mounted on the vehicle.

    (a) grounding electrode
    (b) grounded conductor
    (c) ungrounded conductor
    (d) equipment grounding conductor

37. High-impedance grounded neutral systems shall be permitted for three-phase ac systems of 480 volts to 1,000 volts where _____.

    (a) the conditions of maintenance ensure that only qualified persons service the installation
    (b) ground detectors are installed on the system
    (c) line-to-neutral loads are not served
    (d) all of these

PART III. GROUNDING ELECTRODE SYSTEM AND GROUNDING ELECTRODE CONDUCTOR

250.50 Grounding Electrode System. Any grounding electrode described in 250.52(A)(1) through (A)(8) that is present at a building or structure must be bonded together to form the grounding electrode system. **Figure 250–92**

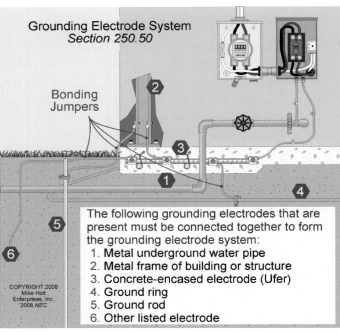

Grounding Electrode System
Section 250.50

Bonding Jumpers

The following grounding electrodes that are present must be connected together to form the grounding electrode system:
1. Metal underground water pipe
2. Metal frame of building or structure
3. Concrete-encased electrode (Ufer)
4. Ground ring
5. Ground rod
6. Other listed electrode

COPYRIGHT 2008 Mike Holt Enterprises, Inc. 2008 NEC

Figure 250–92

- Underground metal water pipe [250.52(A)(1)]
- Metal frame of the building or structure [250.52(A)(2)]
- Concrete-encased electrode [250.52(A)(3)]
- Ground ring [250.52(A)(4)]
- Ground rod [250.52(A)(5)]
- Other listed electrodes [250.52(A)(6)]
- Grounding plate [250.52(A)(7)]

- Metal underground systems, piping systems, or underground tanks [250.52(A)(8)].

Exception: Concrete-encased electrodes are not required for existing buildings or structures where the conductive steel reinforcing bars aren't accessible without chipping up the concrete. **Figure 250–93**

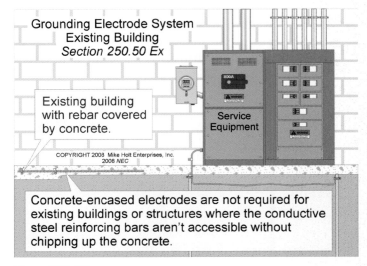

Grounding Electrode System
Existing Building
Section 250.50 Ex

Existing building with rebar covered by concrete.

Service Equipment

COPYRIGHT 2008 Mike Holt Enterprises, Inc. 2008 NEC

Concrete-encased electrodes are not required for existing buildings or structures where the conductive steel reinforcing bars aren't accessible without chipping up the concrete.

Figure 250–93

250.52 Grounding Electrode Types.

(A) Electrodes Permitted for Grounding.

(1) Underground Metal Water Pipe Electrode. Underground metal water pipe in direct contact with earth for 10 ft or more can serve as a grounding electrode. **Figure 250–94**

> **Author's Comment:** The grounding electrode conductor to the water pipe electrode must be sized in accordance with Table 250.66.

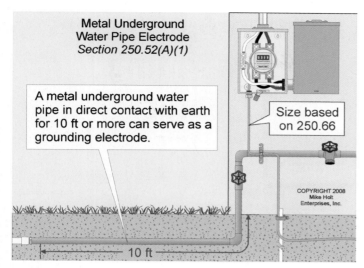

Figure 250–94

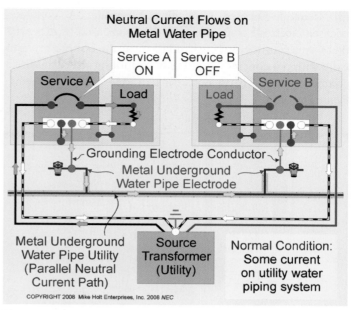

Figure 250–95

An underground metal water pipe electrode that may be interrupted, such as with a water meter, must be made electrically continuous with a bonding jumper sized according to 250.66, based on the area of the ungrounded service conductors [250.68(B)].

Interior metal water piping located more than 5 ft from the point of entrance to the building or structure can't be used to interconnect electrodes that are part of the grounding electrode system.

Exception: In industrial, institutional, and commercial buildings where conditions of maintenance and supervision ensure only qualified persons service the installation, the entire length of the metal water piping system can be used for grounding purposes, provided the entire length, other than short sections passing through walls, floors, or ceilings, is exposed.

Author's Comment: Controversy about using metal underground water supply piping as a grounding electrode has existed since the early 1900s. The water industry believes that neutral current flowing on water piping corrodes the metal. For more information, contact the American Water Works Association about their report *Effects of Electrical Grounding on Pipe Integrity and Shock Hazard*, Catalog No. 90702, 1-800-926-7337. **Figure 250–95**

(2) Metal Building Frame Electrode. The metal frame of a building or structure can serve as a grounding electrode when it meets one of the following conditions:

(1) A single structural metal member (10 ft or more) is in direct contact with the earth or encased in concrete that is in direct contact with the earth,

(2) The structural metal frame is connected to a concrete-encased electrode as provided in 250.52(A)(3) or a ground ring as provided in 250.52(A)(4), **Figure 250–96**

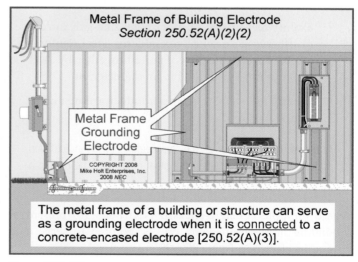

Figure 250–96

(3) The he structural metal frame is connected to a rod electrode [250.52(A)(5)] that has a contact resistance of 25 ohms or less [250.56], or

(4) A single structural metal member is connected to the earth by other approved means.

(3) Concrete-Encased Grounding Electrode. A concrete-encased electrode is an electrode that is encased by at least 2 in. of concrete, located horizontally near the bottom or vertically within a concrete foundation or footing that is in direct contact with the earth consisting of one of the following: **Figure 250–97**

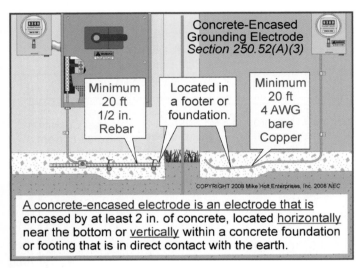

A concrete-encased electrode is an electrode that is encased by at least 2 in. of concrete, located horizontally near the bottom or vertically within a concrete foundation or footing that is in direct contact with the earth.

Figure 250–97

- Twenty feet of one or more bare, zinc-galvanized, or otherwise electrically conductive steel reinforcing bars mechanically connected together by steel tie wires not less than ½ in. in diameter, or

- Twenty feet of bare copper conductor not smaller than 4 AWG.

Author's Comment: If a moisture/vapor barrier is installed under a concrete footer, then the steel rebar can't be part of a concrete-encased electrode.

Where multiple concrete-encased electrodes are present at a building or structure, only one is required to serve as a grounding electrode. **Figure 250–98**

Author's Comments:

- The grounding electrode conductor to a concrete-encased grounding electrode isn't required to be larger than 4 AWG copper [250.66(B)].

- The concrete-encased grounding electrode is also called a "Ufer Ground," named after a consultant working for the US Army during World War II. The technique Mr. Ufer came up

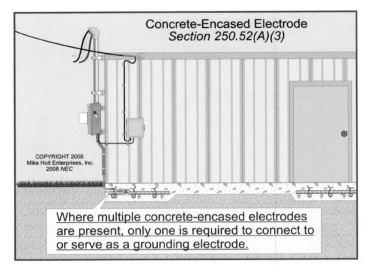

Where multiple concrete-encased electrodes are present, only one is required to connect to or serve as a grounding electrode.

Figure 250–98

with was necessary because the site needing grounding had no underground water table and little rainfall. The desert site was a series of bomb storage vaults in the area of Flagstaff, Arizona. This type of grounding electrode generally offers the lowest ground resistance for the cost.

(4) Ground Ring Electrode. A bare copper conductor, not smaller than 2 AWG buried in the earth encircling a building or structure, can serve as a grounding electrode. **Figure 250–99**

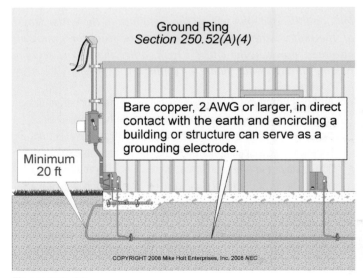

Bare copper, 2 AWG or larger, in direct contact with the earth and encircling a building or structure can serve as a grounding electrode.

Figure 250–99

Author's Comment: The ground ring must be buried not less than 30 in. [250.53(F)], and the grounding electrode conductor to a ground ring isn't required to be larger than the ground ring conductor size [250.66(C)].

(5) Ground Rod Electrode. Ground rod electrodes must not be less than 8 ft in length in contact with the earth [250.53(G)].

(b) Unlisted ground rods of stainless steel, copper coated steel, or zinc coated steel must have a diameter of at least ⅝ in. and listed ground rods must have a diameter of at least ½ in. **Figure 250–100**

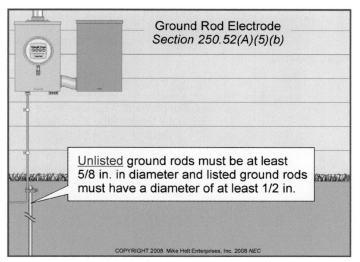

Ground Rod Electrode
Section 250.52(A)(5)(b)

Unlisted ground rods must be at least 5/8 in. in diameter and listed ground rods must have a diameter of at least 1/2 in.

COPYRIGHT 2008 Mike Holt Enterprises, Inc. 2008 *NEC*

Figure 250–100

Author's Comment:

- The grounding electrode conductor, if it's the sole connection to the ground rod, is not required to be larger than 6 AWG copper [250.66(A)].

- The diameter of a ground rod has an insignificant effect on the contact resistance of a ground rod to the earth. However, larger diameter ground rods (¾ in. and 1 in.) are sometimes installed where mechanical strength is desired, or to compensate for the loss of the electrode's metal due to corrosion.

(6) Listed Electrode. Other listed grounding electrodes.

(7) Ground Plate Electrode. A buried iron or steel plate with not less than ¼ in. of thickness, or a copper metal plate not less than 0.06 in. of thickness, with an exposed surface area not less than 2 sq ft.

(8) Metal Underground Systems Electrode. Metal underground piping systems, underground tanks, and underground metal well casings can serve as a grounding electrode.

Author's Comment: The grounding electrode conductor to the metal underground system must be sized according to Table 250.66.

(B) Not Permitted for Use as Grounding Electrode.

(1) Underground metal gas piping systems. **Figure 250–101**

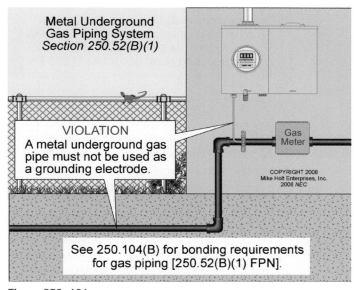

Metal Underground
Gas Piping System
Section 250.52(B)(1)

VIOLATION
A metal underground gas pipe must not be used as a grounding electrode.

Gas Meter

COPYRIGHT 2008
Mike Holt Enterprises, Inc.
2008 *NEC*

See 250.104(B) for bonding requirements for gas piping [250.52(B)(1) FPN].

Figure 250–101

250.53 Grounding Electrode Installation Requirements.

(A) Rod Electrodes. Where practicable, rod electrodes must be embedded below permanent moisture level.

(B) Electrode Spacing. Electrodes for power systems must not be less than 6 ft from any other electrode of another grounding system. Two or more grounding electrodes that are bonded together are considered a single grounding electrode system.

(C) Grounding Electrode Bonding Jumper. Grounding electrode bonding jumpers must be copper when within 18 in. of earth [250.64(A)], be securely fastened to the surface, and be protected if exposed to physical damage [250.64(B)]. The bonding jumper to each electrode must be sized according to 250.66. **Figure 250–102**

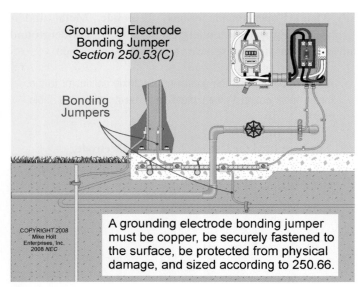

Figure 250–102

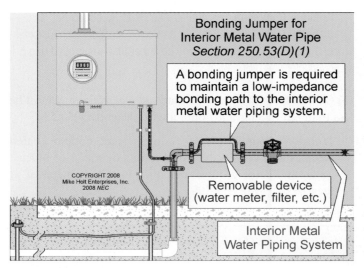

Figure 250–103

The grounding electrode bonding jumpers must terminate by the use of listed pressure connectors, terminal bars, exothermic welding, or other listed means [250.8(A)]. When the termination is encased in concrete or buried, the termination fittings must be listed for this purpose [250.70].

(D) Underground Metal Water Pipe Electrode.

(1) Continuity. The bonding connection to the interior metal water piping system, as required by 250.104(A), must not be dependent on water meters, filtering devices, or similar equipment likely to be disconnected for repairs or replacement. When necessary, a bonding jumper must be installed around insulated joints and equipment likely to be disconnected for repairs or replacement to assist in clearing and removing dangerous voltage on metal parts due to a ground fault [250.68(B)]. **Figure 250–103**

(2) Underground Metal Water Pipe Supplemental Electrode Required. When an underground metal water pipe grounding electrode is present [250.52(A)(1)], it must be supplemented by one of the following electrodes:

- Metal frame of the building or structure electrode [250.52(A)(2)]
- Concrete-encased electrode [250.52(A)(3)]
  Figure 250–104
- Ground ring electrode [250.52(A)(4)]

Where none of the above electrodes are present, one of the following electrodes must be installed to supplement the water pipe electrode:

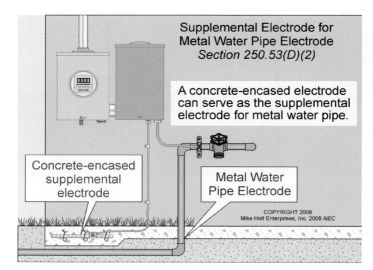

Figure 250–104

- Ground rod electrode meeting the requirements of 250.56 [250.52(A)(5)]
- Other listed electrodes [250.52(A)(6)]
- Metal underground systems, piping systems, or underground tanks [250.52(A)(8)]

The termination of the supplemental grounding electrode conductor must be to one of the following locations: **Figure 250–105**

- Grounding electrode conductor
- Service neutral conductor
- Metal service raceway
- Service equipment enclosure

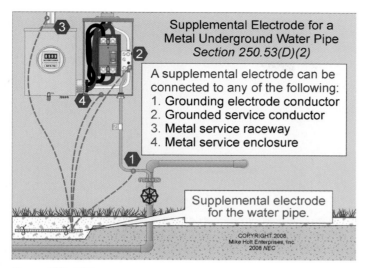

Figure 250–105

(E) Supplemental Ground Rod Electrode. The grounding electrode conductor to a ground rod that serves as a supplemental electrode isn't required to be larger than 6 AWG copper.

(F) Ground Ring. A ground ring encircling the building or structure, consisting of at least 20 ft of bare copper conductor not smaller than 2 AWG, must be buried not less than 30 in. [250.52(A)(4)]. **Figure 250–106**

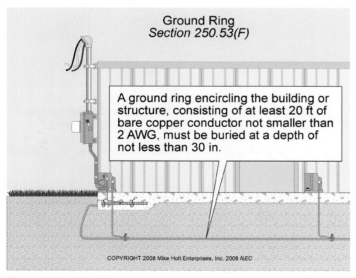

Figure 250–106

(G) Ground Rod Electrodes. Ground rod electrodes must be installed so that not less than 8 ft of length is in contact with the soil. Where rock bottom is encountered, the ground rod must be driven at an angle not to exceed 45 degrees from vertical. If rock bottom is encountered at an angle up to 45 degrees from vertical, the ground rod can be buried in a minimum 30 in. deep trench. **Figure 250–107**

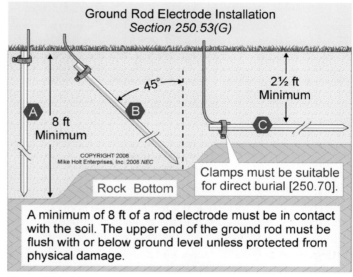

Figure 250–107

The upper end of the ground rod must be flush with or underground unless the grounding electrode conductor attachment is protected against physical damage as specified in 250.10.

> **Author's Comment:** When the grounding electrode attachment fitting is located underground, it must be listed for direct soil burial [250.68(A) Ex 1 and 250.70].

250.54 Auxiliary Grounding Electrodes. Auxiliary electrodes can be connected to the circuit equipment grounding conductor. They are not required to be bonded to the building or structure grounding electrode system, the grounding conductor to the electrode is not required to be sized to 250.66, and its contact resistance to the earth is not required to comply with the 25 ohm requirement of 250.56. **Figures 250–108** and **250–109**

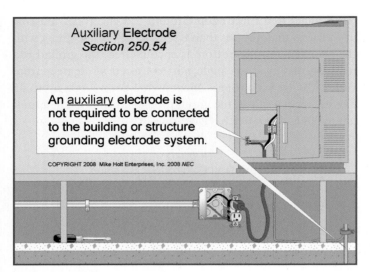

Figure 250–108

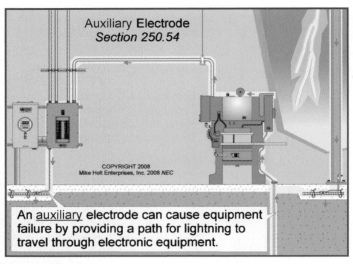

Figure 250–110

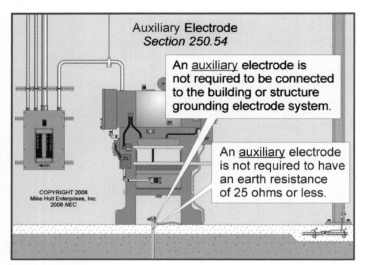

Figure 250–109

CAUTION: *An auxiliary electrode typically serves no useful purpose, and in some cases it may actually cause equipment failures by providing a path for lightning to travel through electronic equipment.* **Figure 250–110**

The earth must not be used as the effective ground-fault current path required by 250.4(A)(5).

DANGER: *Because the contact resistance of an electrode to the earth is so great, very little fault current returns to the power supply if the earth is the only fault current return path. Result—the circuit overcurrent device will not open and clear the ground fault, and all metal parts associated with the electrical installation, metal piping, and structural building steel will become and remain energized.*

250.56 Contact Resistance of Ground Rod to the Earth. When the grounding electrode system consists of a single ground rod having a contact resistance to the earth of over 25 ohms, it must be augmented with an additional electrode located not less than 6 ft away. **Figure 250–111**

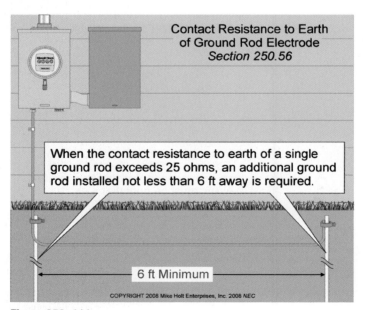

Figure 250–111

Author's Comment: If the contact resistance of two ground rods to the earth exceeds 25 ohms, no additional electrodes are required.

SPECIAL SECTION

Measuring the Ground Resistance

A ground resistance clamp meter, or a three-point fall of potential ground resistance meter, can be used to measure the contact resistance of a grounding electrode to the earth.

Ground Clamp Meter. The ground resistance clamp meter measures the contact resistance of the grounding system to the earth by injecting a high-frequency signal via the service neutral conductor to the utility ground, and then measuring the strength of the return signal through the earth to the grounding electrode being measured. **Figure 250–112**

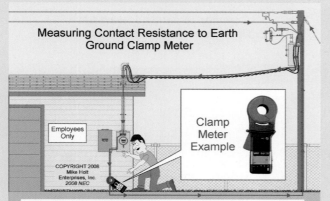

The clamp meter measures the contact resistance to earth of the grounding electrode system by injecting a high-frequency signal to the utility ground, then measuring the strength of the return signal.

Figure 250–112

Fall of Potential Ground Resistance Meter. The three-point fall of potential ground resistance meter determines the contact resistance of a single grounding electrode to the earth by using Ohm's Law: R=E/I.

This meter divides the voltage difference between the electrode to be measured and a driven potential test stake (P) by the current flowing between the electrode to be measured and a driven current test stake (C). The test stakes are typically made of ¼ in. diameter steel rods, 24 in. long, driven two-thirds of their length into earth.

The distance and alignment between the potential and current test stakes, and the electrode, is extremely important to the validity of the earth contact resistance measurements. For an 8 ft ground rod, the accepted practice is to space the current test stake (C) 80 ft from the electrode to be measured.

The potential test stake (P) is positioned in a straight line between the electrode to be measured and the current test stake (C). The potential test stake should be located at approximately 62 percent of the distance the current test stake is located from the electrode. Since the current test stake (C) for an 8 ft ground rod is located 80 ft from the grounding electrode, the potential test stake (P) will be about 50 ft from the electrode to be measured.

Question: If the voltage between the ground rod and the potential test stake (P) is 3V and the current between the ground rod and the current test stake (C) is 0.20A, then the earth contact resistance of the electrode to the earth will be _____. **Figure 250–113**

(a) 5 ohms   (b) 10 ohms   (c) 15 ohms   (d) 25 ohms

Answer: (c) 15 ohms

Resistance = Voltage/Current

E (Voltage) = 3V
I (Current) = 0.20A

R = E/I

Resistance = 3V/0.20A
Resistance = 15 ohms

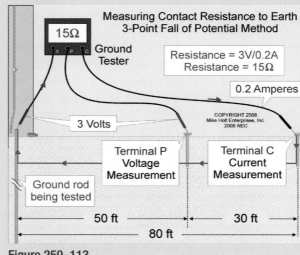

Figure 250–113

Author's Comment: The three-point fall of potential meter can only be used to measure the contact resistance of one electrode to the earth at a time, and this electrode must be independent and not connected to any part of the electrical system. The contact resistance of two electrodes bonded together must not be measured until they have been separated. The contact resistance of two separate electrodes to the earth is calculated as if they are two resistors connected in parallel.

Soil Resistivity

The earth's ground resistance is directly impacted by soil resistivity, which varies throughout the world. Soil resistivity is influenced by electrolytes, which consist of moisture, minerals, and dissolved salts. Because soil resistivity changes with moisture content, the resistance of any grounding system varies with the seasons of the year. Since moisture is stable at greater distances below the surface of the earth, grounding systems are generally more effective if the grounding electrode can reach the water table. In addition, having the grounding electrode below the frost line helps to ensure less deviation in the system's contact resistance to the earth year round.

The contact resistance to the earth can be lowered by chemically treating the earth around the grounding electrodes with electrolytes designed for this purpose.

250.58 Common Grounding Electrode. Where separate services, feeders, or branch circuits supply a building, the same grounding electrode must be used. **Figure 250–114**

Author's Comment: Metal parts of the electrical installation are grounded (connected to the earth) to reduce induced voltage on the metal parts from lightning so as to prevent fires from a surface arc within the building or structure. Grounding electrical equipment doesn't serve the purpose of providing a low-impedance fault current path to open the circuit overcurrent device in the event of a ground fault.

CAUTION: *Potentially dangerous objectionable neutral current flows on the metal parts when multiple service disconnecting means are connected to the same electrode. This is because neutral current from each service can return to the utility via the common grounding electrode and its conductors. This is especially a problem if a service neutral conductor is opened.* Figure 250–115

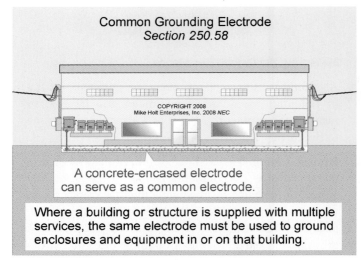

Figure 250–114

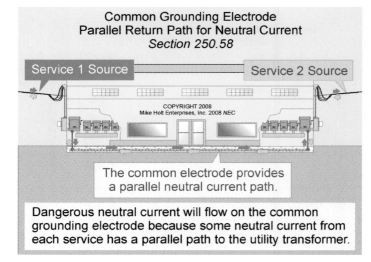

Figure 250–115

250.60 Lightning Protection Electrode. The grounding electrode for a lightning protection system must not be used as the required grounding electrode system for the buildings or structures. **Figure 250–116**

Author's Comment: Where a lightning protection system is installed, the lightning protection system must be bonded to the building or structure grounding electrode system [250.106]. **Figure 250–117**

250.62 Grounding Electrode Conductor. The grounding electrode conductor must be solid or stranded, insulated or bare, and it must be copper if within 18 in. of the earth [250.64(A)]. **Figure 250–118**

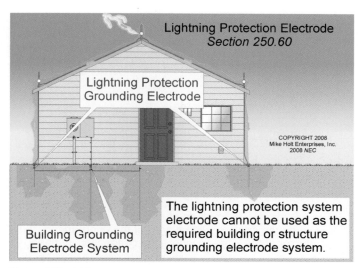

Figure 250–116

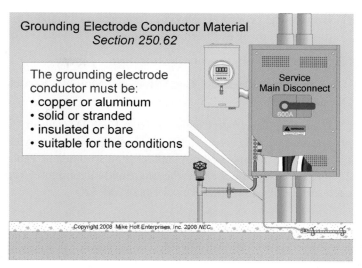

Figure 250–118

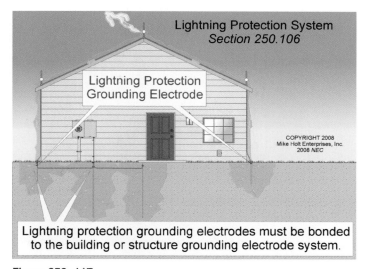

Figure 250–117

250.64 Grounding Electrode Conductor Installation.

Grounding electrode conductors must be installed as specified in (A) through (F).

(A) Aluminum Conductors. Aluminum grounding electrode conductors must not be within 18 in. of earth.

(B) Conductor Protection. Where run exposed, grounding electrode conductors must be protected where subject to physical damage, and grounding electrode conductors 6 AWG copper and larger can be run exposed along the surface of the building if securely fastened and not subject to physical damage.

Grounding electrode conductors sized 8 AWG must be installed in rigid metal conduit, intermediate metal conduit, PVC conduit, or electrical metallic tubing.

> **Author's Comment:** A ferrous metal raceway containing an grounding electrode conductor must be made electrically continuous by bonding each end of the ferrous metal raceway to the grounding electrode conductor [250.64(E)]. So it's best to use PVC conduit.

(C) Continuous Run. The grounding electrode conductor is not permitted to be spliced except as permitted in (1) or (2): Figure 250–119

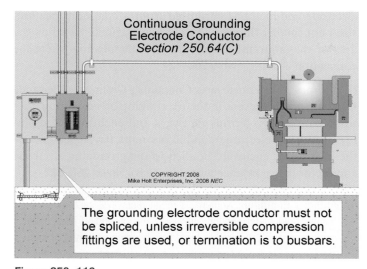

Figure 250–119

(1) Irreversible compression-type connectors listed for grounding, or by exothermic welding.

(2) Sections of busbars connected together.

(D) Grounding Electrode Conductor for Multiple Service Disconnects.

(2) Grounding Electrode Conductor from Each Disconnect. A grounding electrode conductor is permitted from each service disconnecting means sized not smaller than specified in Table 250.66, based on the area of the ungrounded conductor for each service disconnecting means. **Figure 250–120**

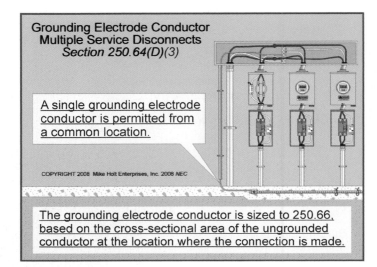

Figure 250–121

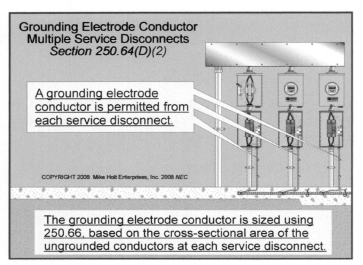

Figure 250–120

(3) One Grounding Electrode Conductor. A single grounding electrode conductor is permitted from a common location, sized not smaller than specified in Table 250.66, based on the area of the ungrounded conductor at the location where the connection is made. **Figure 250–121**

(E) Ferrous Metal Enclosures Containing Grounding Electrode Conductors. To prevent inductive choking of grounding electrode conductors; ferrous raceways and enclosures containing grounding electrode conductors must have each end of the raceway or enclosure bonded to the grounding electrode conductor in accordance with 250.92(B). **Figure 250–122**

> **Author's Comment:** Nonferrous metal raceways, such as aluminum rigid metal conduit, enclosing the grounding electrode conductor aren't required to meet the "bonding each end of the raceway to the grounding electrode conductor" provisions of this section.

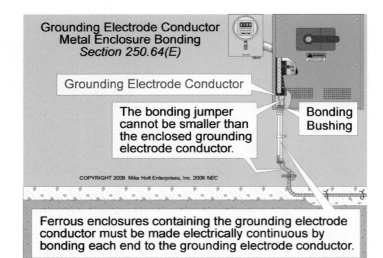

Figure 250–122

> **CAUTION:** *The effectiveness of a grounding electrode is significantly reduced if a ferrous metal raceway containing a grounding electrode conductor isn't bonded to the ferrous metal raceway at both ends. This is because a single conductor carrying high-frequency induced lightning current in a ferrous raceway causes the raceway to act as an inductor, which severely limits (chokes) the current flow through the grounding electrode conductor. ANSI/IEEE 142, Recommended Practice for Grounding of Industrial and Commercial Power Systems (Green Book) states: "An inductive choke can reduce the current flow by 97 percent."*

Author's Comment: To save a lot of time and effort, run the grounding electrode conductor exposed if it's not subject to physical damage [250.64(B)], or enclose it in PVC conduit suitable for the application [352.10(F)].

(F) Termination to Grounding Electrode.

(1) Single Grounding Electrode Conductor. A single grounding electrode conductor is permitted to terminate to any grounding electrode of the grounding electrode system. **Figure 250–123**

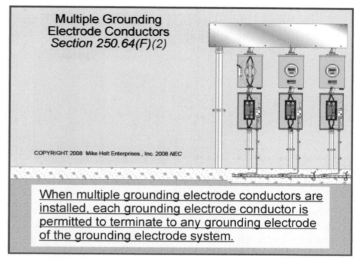

Multiple Grounding Electrode Conductors
Section 250.64(F)(2)

COPYRIGHT 2008 Mike Holt Enterprises , Inc. 2008 NEC

When multiple grounding electrode conductors are installed, each grounding electrode conductor is permitted to terminate to any grounding electrode of the grounding electrode system.

Figure 250–124

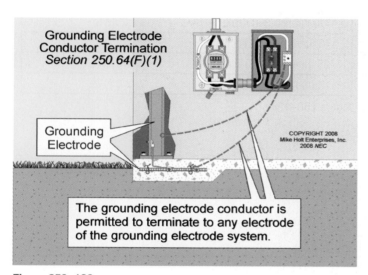

Grounding Electrode Conductor Termination
Section 250.64(F)(1)

Grounding Electrode

COPYRIGHT 2008 Mike Holt Enterprises, Inc. 2008 NEC

The grounding electrode conductor is permitted to terminate to any electrode of the grounding electrode system.

Figure 250–123

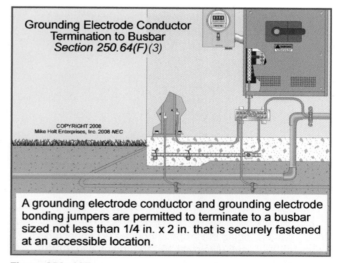

Grounding Electrode Conductor Termination to Busbar
Section 250.64(F)(3)

COPYRIGHT 2008 Mike Holt Enterprises, Inc. 2008 NEC

A grounding electrode conductor and grounding electrode bonding jumpers are permitted to terminate to a busbar sized not less than 1/4 in. x 2 in. that is securely fastened at an accessible location.

Figure 250–125

(2) Multiple Grounding Electrode Conductors. When multiple grounding electrode conductors are installed [250.64(D)(2)], each grounding electrode conductor is permitted to terminate to any grounding electrode of the grounding electrode system. **Figure 250–124**

(3) Termination to Busbar. A grounding electrode conductor and grounding electrode bonding jumpers are permitted to terminate to a busbar sized not less than ¼ in. × 2 in. that is securely fastened at an accessible location. The terminations to the busbar must be made by a listed connector or by exothermic welding. **Figure 250–125**

250.66 Sizing Grounding Electrode Conductor. Except as permitted in (A) through (C), a grounding electrode conductor must be sized in accordance with Table 250.66.

(A) Ground Rod. Where the grounding electrode conductor is connected to a ground rod as permitted in 250.52(A)(5), that portion of the grounding electrode conductor that is the sole connection to the ground rod isn't required to be larger than 6 AWG copper. **Figure 250–126**

(B) Concrete-Encased Grounding Electrode. Where the grounding electrode conductor is connected to a concrete-encased electrode, that portion of the grounding electrode conductor that is the sole connection to the concrete-encased electrode isn't required to be larger than 4 AWG copper. **Figure 250–127**

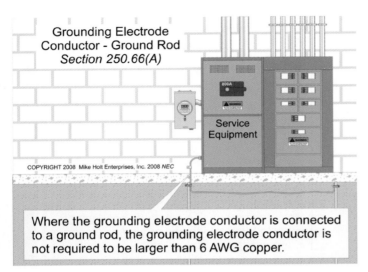

Grounding Electrode
Conductor - Ground Rod
Section 250.66(A)

COPYRIGHT 2008 Mike Holt Enterprises, Inc. 2008 *NEC*

Where the grounding electrode conductor is connected to a ground rod, the grounding electrode conductor is not required to be larger than 6 AWG copper.

Figure 250–126

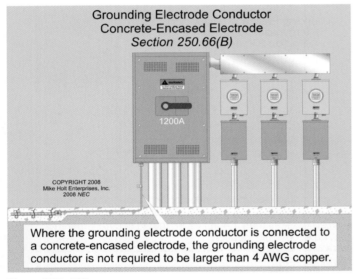

Grounding Electrode Conductor
Concrete-Encased Electrode
Section 250.66(B)

1200A

COPYRIGHT 2008
Mike Holt Enterprises, Inc.
2008 *NEC*

Where the grounding electrode conductor is connected to a concrete-encased electrode, the grounding electrode conductor is not required to be larger than 4 AWG copper.

Figure 250–127

(C) Ground Ring. Where the grounding electrode conductor is connected to a ground ring, that portion of the conductor that is the sole connection to the ground ring isn't required to be larger than the conductor used for the ground ring.

Author's Comment: A ground ring encircling the building or structure in direct contact with earth must consist of at least 20 ft of bare copper conductor not smaller than 2 AWG [250.52(A)(4)]. See 250.53(F) for the installation requirements for a ground ring.

Author's Comment: Table 250.66 is used to size the grounding electrode conductor when the conditions of 250.66(A), (B), or (C) do not apply.

Table 250.66 Sizing Grounding Electrode Conductor	
Conductor or Area of Parallel Conductors	**Copper Grounding Electrode Conductor**
12 through 2 AWG	8 AWG
1 or 1/0 AWG	6 AWG
2/0 or 3/0 AWG	4 AWG
4/0 through 350 kcmil	2 AWG
400 through 600 kcmil	1/0 AWG
700 through 1,100 kcmil	2/0 AWG
1,200 kcmil and larger	3/0 AWG

Figure 250–128

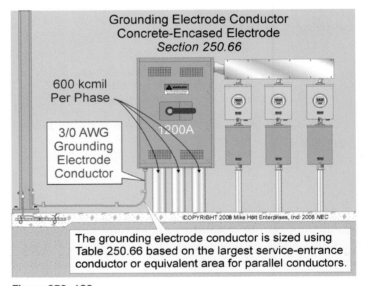

Grounding Electrode Conductor
Concrete-Encased Electrode
Section 250.66

600 kcmil
Per Phase

3/0 AWG
Grounding
Electrode
Conductor

1200A

COPYRIGHT 2008 Mike Holt Enterprises, Inc. 2008 *NEC*

The grounding electrode conductor is sized using Table 250.66 based on the largest service-entrance conductor or equivalent area for parallel conductors.

Figure 250–128

250.68 Termination to the Grounding Electrode.

(A) Accessibility. The mechanical elements used to terminate a grounding electrode conductor or bonding jumper to a grounding electrode must be accessible. **Figure 250–129**

Exception No. 1: The termination is not required to be accessible if the termination to the electrode is encased in concrete or buried in the earth. **Figure 250–130**

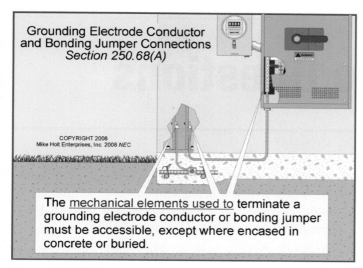

Figure 250–129

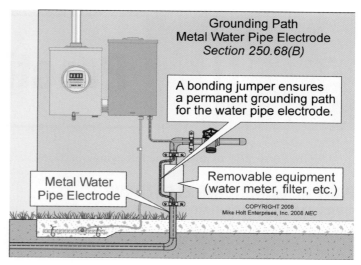

Figure 250–131

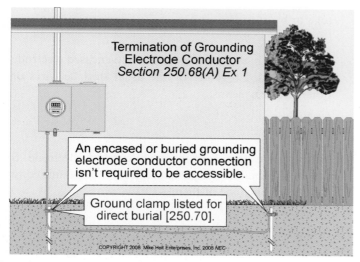

Figure 250–130

Author's Comment: Where the grounding electrode attachment fitting is encased in concrete or buried in the earth, it must be listed for direct soil burial or concrete encasement [250.70].

Exception No. 2: Exothermic or irreversible compression connections, together with the mechanical means used to attach to fireproofed structural metal, are not required to be accessible.

(B) Integrity of Underground Metal Water Pipe Electrode. A bonding jumper must be installed around insulated joints and equipment likely to be disconnected for repairs or replacement for an underground water metal piping system used as a grounding electrode. The bonding jumper must be of sufficient length to allow the removal of such equipment while retaining the integrity of the grounding path. **Figure 250–131**

250.70 Grounding Electrode Conductor Termination Fittings. The grounding electrode conductor must terminate to the grounding electrode by exothermic welding, listed lugs, listed pressure connectors, listed clamps, or other listed means. In addition, fittings terminating to a grounding electrode must be listed for the materials of the grounding electrode.

When the termination to a grounding electrode is encased in concrete or buried in the earth, the termination fitting must be listed for direct soil burial or concrete encasement. No more than one conductor can terminate on a single clamp or fitting unless the clamp or fitting is listed for multiple connections. **Figure 250–132**

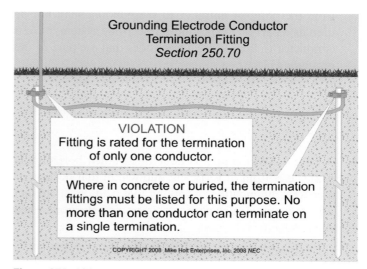

Figure 250–132

PART III Practice Questions

Use the 2008 *NEC* to answer the following questions.

PART III. GROUNDING ELECTRODE SYSTEM AND GROUNDING ELECTRODE CONDUCTOR PRACTICE QUESTIONS

1. Concrete-encased electrodes of _____ shall not be required to be part of the grounding electrode system where the steel reinforcing bars or rods aren't accessible for use without disturbing the concrete.

   (a) hazardous (classified) locations
   (b) health care facilities
   (c) existing buildings or structures
   (d) agricultural buildings with equipotential planes

2. In order for a metal underground water pipe to be used as a grounding electrode, it shall be in direct contact with the earth for _____ .

   (a) 5 ft
   (b) 10 ft or more
   (c) less than 10 ft
   (d) 20 ft or more

3. Interior metal water piping located more than _____ from the point of entrance to the building shall not be used as a part of the grounding electrode system, or as a conductor to interconnect electrodes that are part of the grounding electrode system.

   (a) 2 ft
   (b) 4 ft
   (c) 5 ft
   (d) 6 ft

4. A bare 4 AWG copper conductor installed horizontally near the bottom or vertically, and within that portion of a concrete foundation or footing that is in direct contact with the earth can be used as a grounding electrode when the conductor is at least _____ in length.

   (a) 10 ft
   (b) 15 ft
   (c) 20 ft
   (d) 25 ft

5. An electrode encased by at least 2 in. of concrete, located horizontally near the bottom or vertically and within that portion of a concrete foundation or footing that is in direct contact with the earth, shall be permitted as a grounding electrode when it consists of _____ .

   (a) at least 20 ft of ½ in. or larger steel reinforcing bars or rods
   (b) at least 20 ft of bare copper conductor of 4 AWG or larger
   (c) a or b
   (d) none of these

6. Reinforcing bars for use as a concrete-encased electrode can be bonded together by the usual steel tie wires or other effective means.

   (a) True
   (b) False

7. Where more than one concrete-encased electrode is present at a building or structure, it shall be permitted to connect to only one of them.

   (a) True
   (b) False

8. A ground ring encircling the building or structure can be used as a grounding electrode when _____ .

   (a) the ring is in direct contact with the earth
   (b) the ring consists of at least 20 ft of bare conductor
   (c) the bare copper conductor is not smaller than 2 AWG
   (d) all of these

9. Grounding electrodes that are driven rods require a minimum of _____ in contact with the soil.

   (a) 6 ft
   (b) 8 ft
   (c) 10 ft
   (d) 12 ft

10. Grounding electrodes of the rod type less than 5/8 in. in diameter shall be listed and shall not be less than _____ in diameter.

(a) ½ in.
(b) ¾ in.
(c) 1 in.
(d) 1¼ in.

11. Local metal underground systems or structures such as _____ are permitted to serve as grounding electrodes.

(a) piping systems
(b) underground tanks
(c) underground metal well casings
(d) all of these

12. _____ shall not be used as grounding electrodes.

(a) Underground gas piping systems
(b) Aluminum
(c) Metal well casings
(d) a and b

13. Where practicable, rod, pipe, and plate electrodes shall be installed _____.

(a) directly below the electrical meter
(b) on the north side of the building
(c) below permanent moisture level
(d) all of these

14. Two or more grounding electrodes bonded together are considered a single grounding electrode system.

(a) True
(b) False

15. Where a metal underground water pipe is used as a grounding electrode, the continuity of the grounding path or the bonding connection to interior piping shall not rely on _____ and similar equipment.

(a) bonding jumpers
(b) water meters or filtering devices
(c) grounding clamps
(d) all of these

16. Where the supplemental electrode is a rod, that portion of the bonding jumper that is the sole connection to the supplemental grounding electrode shall not be required to be larger than _____ AWG.

(a) 8
(b) 6
(c) 4
(d) 1

17. When a ground ring is used as a grounding electrode, it shall be buried at a depth below the earth's surface of not less than _____.

(a) 18 in.
(b) 24 in.
(c) 30 in.
(d) 8 ft

18. Ground rod electrodes shall be installed so that at least _____ of the length is in contact with the soil.

(a) 5 ft
(b) 8 ft
(c) one-half
(d) 80 percent

19. The upper end of a ground rod electrode shall be _____ ground level unless the aboveground end and the grounding electrode conductor attachment are protected against physical damage.

(a) above
(b) flush with
(c) below
(d) b or c

20. Where rock bottom is encountered when driving a ground rod at an angle up to 45 degrees, the electrode can be buried in a trench that is at least _____ deep.

(a) 18 in.
(b) 30 in.
(c) 4 ft
(d) 8 ft

21. Auxiliary grounding electrodes can be connected to the _____.

(a) equipment grounding conductor
(b) grounded conductor
(c) a and b
(d) none of these

22. The auxiliary electrode shall not be used as an effective ground-fault current path.

(a) True
(b) False

23. Where the resistance-to-ground of 25 ohms or less is not achieved for a single rod electrode, _____.

    (a) other means besides electrodes shall be used in order to provide grounding
    (b) the single rod electrode shall be augmented by one additional electrode
    (c) no additional electrodes are required
    (d) none of these

24. Buildings or structures supplied by multiple services or feeders must use the same _____ to ground enclosures and equipment in or on that building.

    (a) service
    (b) disconnect
    (c) electrode system
    (d) any of these

25. The grounding electrode for a lightning protection system can be used as a grounding electrode system for the buildings or structures.

    (a) True
    (b) False

26. Grounding electrode conductors shall be made of a _____ wire.

    (a) solid
    (b) stranded
    (c) insulated or bare
    (d) any of these

27. Where used outside, aluminum or copper-clad aluminum grounding electrode conductors shall not be terminated within _____ of the earth.

    (a) 6 in.
    (b) 12 in.
    (c) 15 in.
    (d) 18 in.

28. Grounding electrode conductors _____ and larger that are not subject to physical damage can be run exposed along the surface of the building construction.

    (a) 10 AWG
    (b) 8 AWG
    (c) 6 AWG
    (d) 4 AWG

29. Grounding electrode conductors smaller than _____ shall be in rigid metal conduit, IMC, PVC conduit, electrical metallic tubing, or cable armor.

    (a) 10 AWG
    (b) 8 AWG
    (c) 6 AWG
    (d) 4 AWG

30. Grounding electrode conductors shall be installed in one continuous length without a splice or joint, unless spliced _____.

    (a) by connecting to a busbar
    (b) by irreversible compression-type connectors listed as grounding and bonding equipment
    (c) by the exothermic welding process
    (d) any of these

31. Ferrous metal enclosures for grounding electrode conductors shall be electrically continuous, from the point of attachment to cabinets or equipment, to the grounding electrode.

    (a) True
    (b) False

32. A service consisting of 12 AWG service-entrance conductors requires a grounding electrode conductor sized no less than _____.

    (a) 10 AWG
    (b) 8 AWG
    (c) 6 AWG
    (d) 4 AWG

33. The largest size grounding electrode conductor required is _____ copper.

    (a) 6 AWG
    (b) 1/0 AWG
    (c) 3/0 AWG
    (d) 250 kcmil

34. What size copper grounding electrode conductor is required for a service that has three sets of 600 kcmil copper conductors per phase?

    (a) 1 AWG
    (b) 1/0 AWG
    (c) 2/0 AWG
    (d) 3/0 AWG

35. In an ac system, the size of the grounding electrode conductor to a concrete-encased electrode shall not be required to be larger than a(n) _____ copper conductor.

(a) 10 AWG
(b) 8 AWG
(c) 6 AWG
(d) 4 AWG

36. Mechanical elements used to terminate a grounding electrode conductor to a grounding electrode shall be accessible.

(a) True
(b) False

37. Grounding electrode conductor connections to a concrete-encased or buried grounding electrode shall be accessible.

(a) True
(b) False

38. Exothermic or irreversible compression connections, together with the mechanical means used to attach to fire-proofed structural metal, shall not be required to be accessible.

(a) True
(b) False

39. When an underground metal water piping system is used as a grounding electrode, bonding shall be provided around insulated joints and around any equipment that is likely to be disconnected for repairs or replacement.

(a) True
(b) False

40. The grounding conductor connection to the grounding electrode shall be made by _____.

(a) listed lugs
(b) exothermic welding
(c) listed pressure connectors
(d) any of these

PART IV. GROUNDING ENCLOSURE, RACEWAY, AND SERVICE CABLE CONNECTIONS

250.86 Other Enclosures. Metal raceways and enclosures containing electrical conductors operating at 50V or more [250.20(A)] must be connected to the circuit equipment grounding conductor. **Figure 250–133**

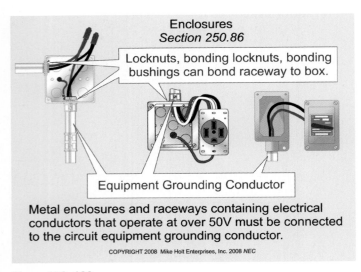

Enclosures
Section 250.86

Locknuts, bonding locknuts, bonding bushings can bond raceway to box.

Equipment Grounding Conductor

Metal enclosures and raceways containing electrical conductors that operate at over 50V must be connected to the circuit equipment grounding conductor.

COPYRIGHT 2008 Mike Holt Enterprises, Inc. 2008 NEC

Figure 250–133

Exception No. 2: Short sections of metal raceways used for the support or physical protection of cables aren't required to be connected to the circuit equipment grounding conductor. **Figure 250–134**

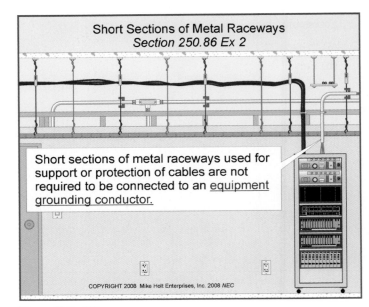

Short Sections of Metal Raceways
Section 250.86 Ex 2

Short sections of metal raceways used for support or protection of cables are not required to be connected to an <u>equipment grounding conductor.</u>

COPYRIGHT 2008 Mike Holt Enterprises, Inc. 2008 NEC

Figure 250–134

PART

IV

Practice Questions

Use the 2008 *NEC* to answer the following questions.

PART IV. ENCLOSURE, RACEWAY, AND SERVICE CABLE CONNECTIONS PRACTICE QUESTIONS

1. Metal enclosures and raceways for other than service conductors shall be connected to the grounded conductor.

    (a) True
    (b) False

2. Short sections of metal enclosures or raceways used to provide support or protection of _____ from physical damage shall not be required to be connected to the equipment grounding conductor.

    (a) conduit
    (b) feeders under 600V
    (c) cable assemblies
    (d) none of these

ARTICLE 250 Grounding and Bonding—Part V. Bonding

PART V. BONDING

250.92 Service Raceways and Enclosures.

(A) Bonding Requirements. The metal parts of equipment indicated in (A)(1) and (A)(2) must be bonded together in accordance with 250.92(B). **Figure 250–135**

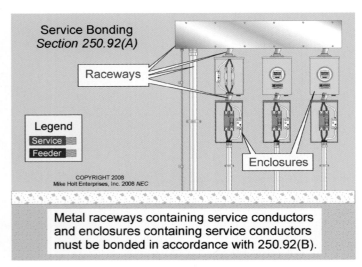

Service Bonding
Section 250.92(A)

Raceways

Legend
Service
Feeder

Enclosures

COPYRIGHT 2008
Mike Holt Enterprises, Inc. 2008 NEC

Metal raceways containing service conductors and enclosures containing service conductors must be bonded in accordance with 250.92(B).

Figure 250–135

(1) Metal raceways containing service conductors.

(2) Metal enclosures containing service conductors.

> **Author's Comment:** Metal raceways or metal enclosures containing feeder and branch-circuit conductors are required to be connected to the circuit equipment grounding conductor in accordance with 250.86. **Figure 250–136**

(B) Methods of Bonding. Metal raceways and metal enclosures containing service conductors must be bonded by one of the following methods:

(1) Neutral Conductor. By bonding the metal parts to the service neutral conductor. **Figure 250–137**

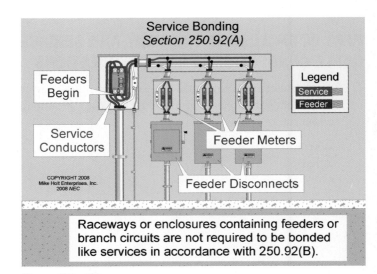

Service Bonding
Section 250.92(A)

Feeders Begin

Legend
Service
Feeder

Service Conductors

Feeder Meters

Feeder Disconnects

COPYRIGHT 2008
Mike Holt Enterprises, Inc.
2008 NEC

Raceways or enclosures containing feeders or branch circuits are not required to be bonded like services in accordance with 250.92(B).

Figure 250–136

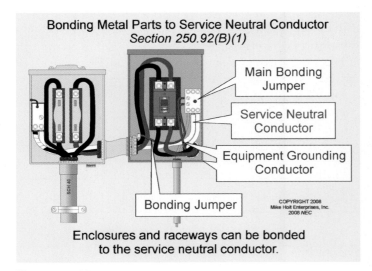

Bonding Metal Parts to Service Neutral Conductor
Section 250.92(B)(1)

Main Bonding Jumper

Service Neutral Conductor

Equipment Grounding Conductor

Bonding Jumper

COPYRIGHT 2008
Mike Holt Enterprises, Inc.
2008 NEC

Enclosures and raceways can be bonded to the service neutral conductor.

Figure 250–137

Author's Comments:

- A main bonding jumper is required to bond the service disconnect to the service neutral conductor [250.24(B) and 250.28].

• At service equipment, the service neutral conductor provides the effective ground-fault current path to the power supply [250.24(C)]; therefore, an equipment grounding conductor isn't required to be installed within PVC conduit containing service-entrance conductors [250.142(A)(1) and 352.60 Ex 2]. **Figure 250–138**

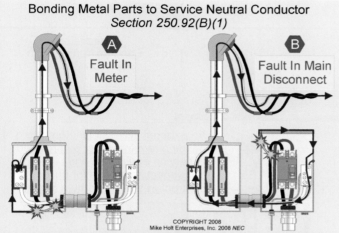

An equipment bonding jumper is not required within rigid nonmetallic conduit because fault current uses the service neutral conductor as the effective ground-fault current path.

Figure 250–138

(2) Threaded Fittings or Entries. By using threaded couplings or threaded entries made up wrenchtight. **Figure 250–139**

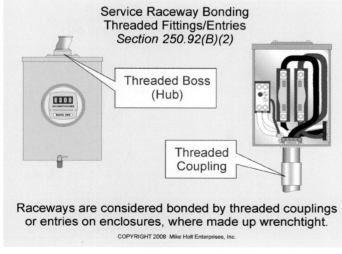

Raceways are considered bonded by threaded couplings or entries on enclosures, where made up wrenchtight.

Figure 250–139

(3) Threadless Fittings. By using threadless raceway couplings and connectors made up tight. **Figure 250–140**

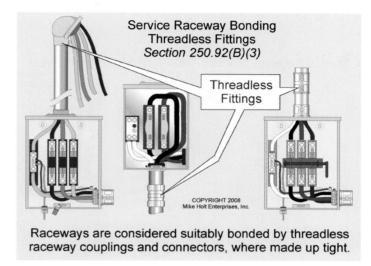

Raceways are considered suitably bonded by threadless raceway couplings and connectors, where made up tight.

Figure 250–140

(4) Bonding Fittings. When a metal service raceway terminates to an enclosure with a ringed knockout, a listed bonding wedge or bushing with a bonding jumper must be used to bond one end of the service raceway to the service neutral conductor. The bonding jumper used for this purpose must be sized in accordance with Table 250.66, based on the area of the service conductors within the raceway [250.92(B)(4) and 250.102(C)]. **Figure 250–141**

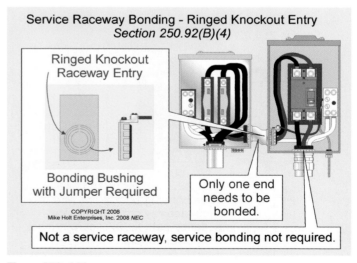

Figure 250–141

Author's Comments:

- When a metal raceway containing service conductors terminates to an enclosure without a ringed knockout, a bonding-type locknut can be used. **Figure 250–142**

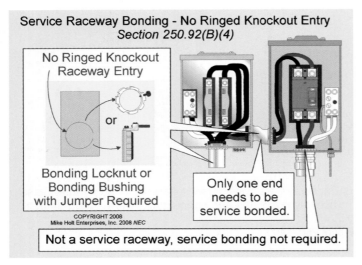

Figure 250–142

- A bonding locknut differs from a standard locknut in that it has a bonding screw with a sharp point that drives into the metal enclosure to ensure a solid connection.
- Bonding one end of a service raceway to the service neutral provides the low-impedance fault current path to the source. **Figure 250–143**

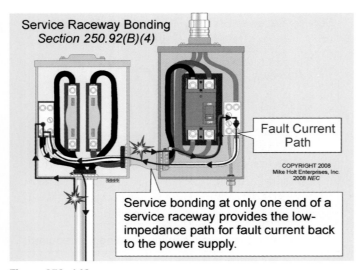

Figure 250–143

250.94 Intersystem Bonding Terminal. An external accessible intersystem bonding terminal for the grounding and bonding of communications systems must be provided at service equipment and disconnecting means for buildings or structures supplied by a feeder. The intersystem bonding terminal must not interfere with the opening of any equipment enclosure and be one of the following: **Figure 250–144**

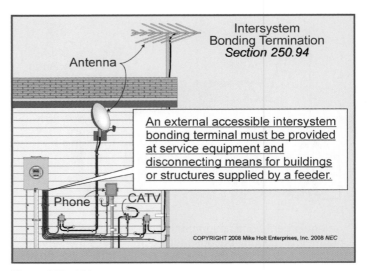

Figure 250–144

(1) Terminals listed for grounding and bonding attached to a meter socket enclosure. **Figure 250–145**

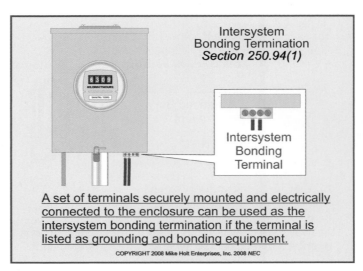

Figure 250–145

(2) Bonding bar connected to the service equipment enclosure or metal service raceway with a minimum 6 AWG copper conductor.

(3) Bonding bar connected to the grounding electrode conductor with a minimum 6 AWG copper conductor. **Figure 250–146**

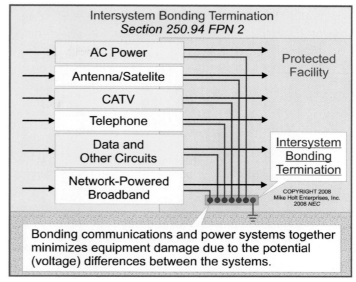

Figure 250–147

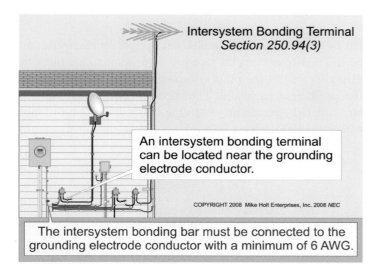

Figure 250–146

Author's Comment: According to Article 100, an intersystem bonding terminal is a device that provides a means to connect communications systems grounding and bonding conductors to the building grounding electrode system.

Exception: At existing buildings or structures, an external accessible means for bonding communications systems together can be by the use of:

(1) Nonflexible metallic raceway,

(2) Grounding electrode conductor, or

(3) Connection approved by the authority having jurisdiction.

FPN No. 2: Communications systems must be bonded to the intersystem bonding terminal in accordance with the following: **Figures 250–147 and 250–148**

* Antennas/Satellite Dishes, 810.15 and 810.21
* CATV, 820.100
* Telephone Circuits, 800.100

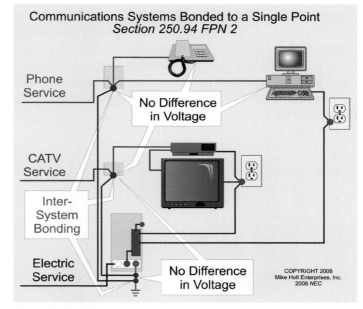

Figure 250–148

Author's Comment: All external communications systems must be connected to the intersystem bonding terminal to minimize the damage to communications systems from induced potential (voltage) differences between the systems from a lightning event. **Figure 250–149**

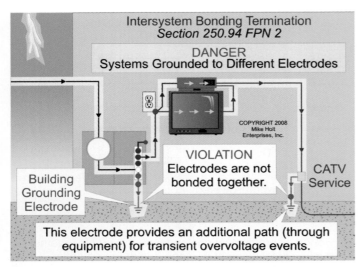

Figure 250–149

250.96 Bonding Other Enclosures.

(A) Maintaining Effective Ground-Fault Current Path. Metal parts intended to serve as equipment grounding conductors including raceways, cables, equipment, and enclosures must be bonded together to ensure they have the capacity to conduct safely any fault current likely to be imposed on them [110.10, 250.4(A)(5), and Note to Table 250.122]. **Figure 250–150**

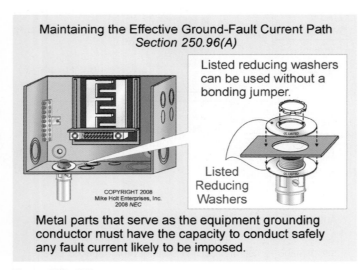

Figure 250–150

Nonconductive coatings such as paint, lacquer, and enamel on equipment must be removed to ensure an effective ground-fault current path, or the termination fittings must be designed so as to make such removal unnecessary [250.12].

> **Author's Comment:** The practice of driving a locknut tight with a screwdriver and pliers is considered sufficient in removing paint and other nonconductive finishes to ensure an effective ground-fault current path.

(B) Isolated Grounding Circuits. An equipment enclosure can be isolated from a metal raceway by a nonmetallic raceway fitting located at the point of attachment of the raceway to the equipment enclosure. The metal raceway must contain an insulated equipment grounding conductor in accordance with 250.146(D). **Figure 250–151**

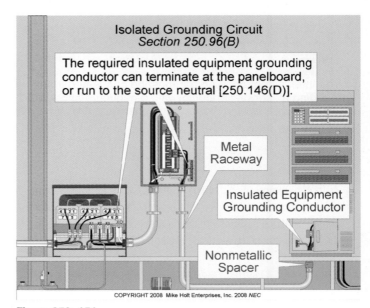

Figure 250–151

250.97 Bonding Metal Parts Containing 277V and 480V Circuits.
Metal raceways or cables containing 277V and/or 480V feeder or branch circuits terminating at ringed knockouts must be bonded to the metal enclosure with a bonding jumper sized in accordance with 250.122, based on the rating of the circuit overcurrent device [250.102(D)]. **Figure 250–152**

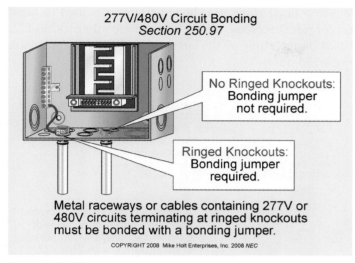

Figure 250–152

Author's Comments:

- Bonding jumpers for raceways and cables containing 277V or 480V circuits are required at ringed knockout terminations to ensure the ground-fault current path has the capacity to safely conduct the maximum ground-fault current likely to be imposed [110.10, 250.4(A)(5), and 250.96(A)].

- Ringed knockouts aren't listed to withstand the heat generated by a 277V ground fault, which generates five times as much heat as a 120V ground fault. **Figure 250–153**

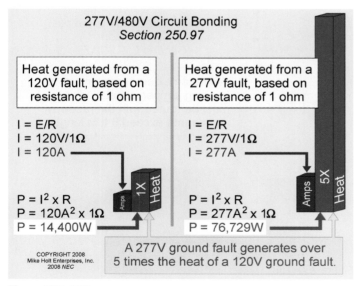

Figure 250–153

Exception: A bonding jumper isn't required where ringed knockouts aren't encountered, knockouts are totally punched out, or where the box is listed to provide a reliable bonding connection. **Figure 250–154**

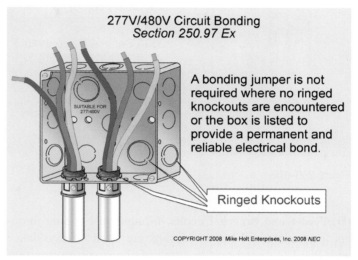

Figure 250–154

250.102 Equipment Bonding Jumpers.

(A) Material. Equipment Bonding jumpers must be copper.

(B) Termination. Equipment bonding jumpers must terminate by listed pressure connectors, terminal bars, exothermic welding, or other listed means [250.8(A)].

(C) Service Equipment. Equipment bonding jumpers for service raceways are sized in accordance with Table 250.66, based on the largest ungrounded conductor within the raceway or cable. Where service conductors are paralleled in two or more raceways or cables, the bonding jumper for each raceway or cable must be sized in accordance with Table 250.66, based on the largest ungrounded conductors within the raceway or cable.

> **Question:** What size equipment bonding jumper is required for a metal raceway containing 700 kcmil service conductors? **Figure 250–155**
>
> (a) 1 AWG   (b) 1/0 AWG   (c) 2/0 AWG   (d) 3/0 AWG
>
> **Answer:** (c) 2/0 AWG [Table 250.66]

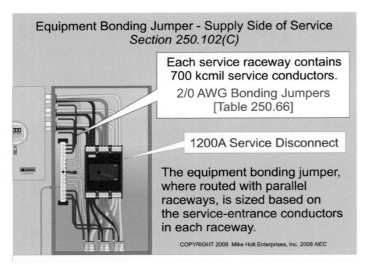

Figure 250–155

(D) Feeder and Branch Circuits. Equipment bonding jumpers for feeders and branch circuits are sized in accordance with 250.122, based on the rating of the circuit overcurrent device.

Question: What size equipment bonding jumper is required for a metal raceway where the circuit conductors are protected by a 1,200A overcurrent device? **Figure 250–156**

(a) 1 AWG   (b) 1/0 AWG (c) 2/0 AWG (d) 3/0 AWG

Answer: (d) 3/0 AWG [Table 250.122]

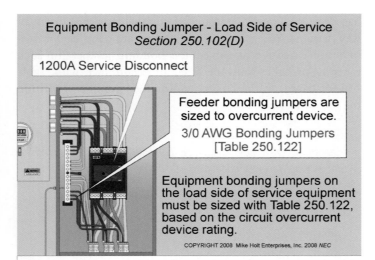

Figure 250–156

Where a single equipment bonding jumper is used to bond two or more raceways, it must be sized according to 250.122, based on the rating of the largest circuit overcurrent device. **Figure 250–157**

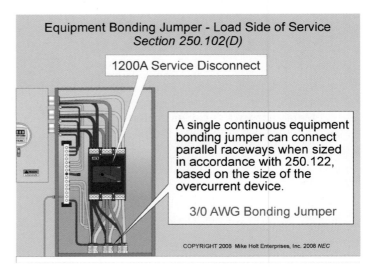

Figure 250–157

(E) Installation. Where the equipment bonding jumper is installed outside of a raceway, its length must not exceed 6 ft and it must be routed with the raceway. **Figure 250–158**

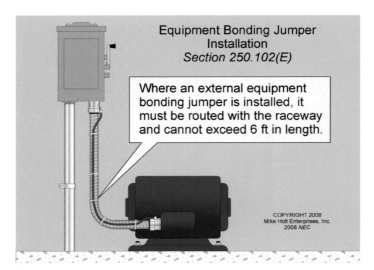

Figure 250–158

250.104 Bonding of Piping Systems and Exposed Structural Metal.

Author's Comment: To remove dangerous voltage on metal parts from a ground fault, electrically conductive metal water piping systems, metal sprinkler piping, metal gas piping, as well as exposed structural steel members likely to become energized, must be connected to an effective ground-fault current path [250.4(A)(4)].

(A) Metal Water Piping System. The metal water piping system must be bonded as required in (A)(1), (A)(2), or (A) (3). The bonding jumper must be copper where within 18 in. of earth [250.64(A)], securely fastened to the surface on which it's mounted [250.64(B)], and adequately protected if exposed to physical damage [250.64(B)]. In addition, all points of attachment must be accessible.

Author's Comment: Bonding isn't required for isolated sections of metal water piping connected to a nonmetallic water piping system. **Figure 250-159**

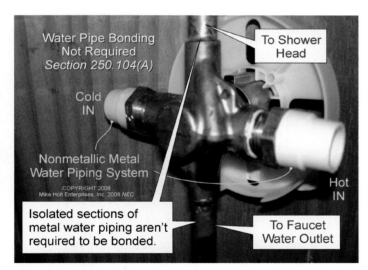

Figure 250–159

(1) Building or Structure Supplied by a Service. The metal water piping system, including the metal sprinkler water piping system of a building or structure supplied with service conductors must be bonded to: **Figure 250–160**

- Service equipment enclosure,
- Service neutral conductor,
- Grounding electrode conductor of sufficient size, or
- The grounding electrode system.

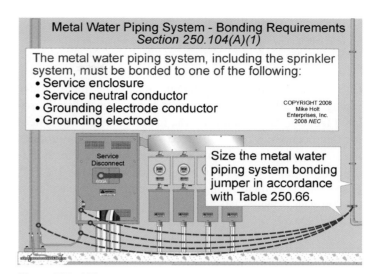

Figure 250–160

The metal water piping system bonding jumper must be sized according to Table 250.66, based on the cross-sectional area of the ungrounded service conductors.

Question: What size bonding jumper is required for the metal water piping system if the 300 kcmil service conductors are paralleled in two raceways? **Figure 250–161**

(a) 6 AWG   (b) 4 AWG   (c) 2 AWG   (d) 1/0 AWG

Answer: (d) 1/0 AWG, based on 600 kcmil conductors, in accordance with Table 250.66

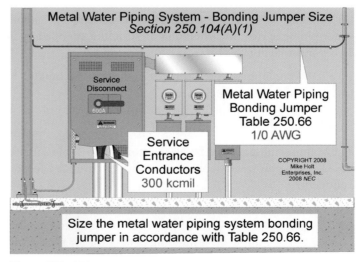

Figure 250–161

Author's Comment: Where hot and cold metal water pipes are electrically connected, only one bonding jumper is required, either to the cold or hot water pipe.

(2) Multiple Occupancy Building. When the metal water piping system in an individual occupancy is metallically isolated from other occupancies, the metal water piping system for that occupancy can be bonded to the equipment grounding terminal of the occupancy's panelboard. The bonding jumper must be sized in accordance with Table 250.122, based on the ampere rating of the occupancy's feeder overcurrent device. **Figure 250–162**

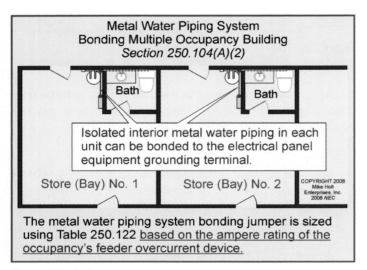

Figure 250–162

(3) Building or Structure Supplied by a Feeder. The metal water piping system of a building or structure supplied by a feeder must be bonded to:

- The equipment grounding terminal of the building disconnect enclosure,
- The feeder equipment grounding conductor, or
- The grounding electrode system.

The bonding jumper must be sized according to Table 250.66, based on the cross-sectional area of the ungrounded feeder conductor.

(B) Other Metal Piping Systems. Metal piping systems such as gas or air that are likely to become energized can be bonded to the equipment grounding conductor for the circuit that may energize the piping. **Figure 250–163**

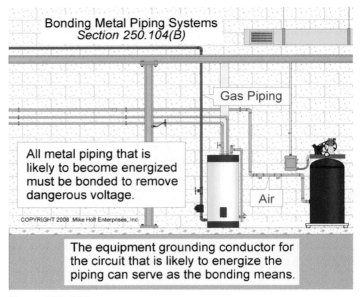

Figure 250–163

(C) Structural Metal. Exposed structural metal that forms a metal building frame that is likely to become energized must be bonded to: **Figure 250–164**

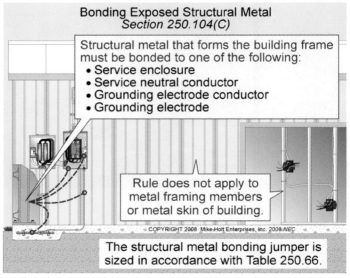

Figure 250–164

- Service equipment enclosure,
- Service neutral conductor,
- Grounding electrode conductor of sufficient size, or
- The grounding electrode system.

Author's Comment: This rule doesn't require the bonding of sheet metal framing members (studs) or the metal skin of a wood frame building.

The bonding jumper must be sized in accordance with Table 250.66, based on the area of ungrounded supply conductors. The bonding jumper must be copper where within 18 in. of earth [250.64(A)], securely fastened to the surface on which it's carried [250.64(B)], and adequately protected if exposed to physical damage [250.64(B)]. In addition, all points of attachment must be accessible.

(D) Separately Derived Systems. Metal water piping systems and structural metal that forms a building frame must be bonded as required in (D)(1) through (D)(3).

(1) Metal Water Pipe. The nearest available point of the metal water piping system in the area served by a separately derived system must be bonded to the neutral point of the separately derived system where the grounding electrode conductor is connected. **Figure 250–165**

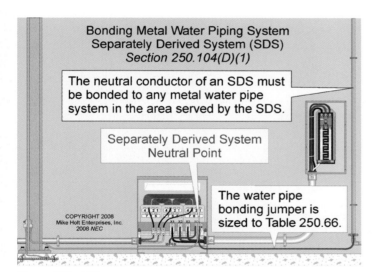

Figure 250–165

The bonding jumper must be sized in accordance with Table 250.66, based on the area of the ungrounded secondary conductors.

Exception No. 2: The metal water piping system is permitted to be bonded to the structural metal building frame if it serves as the grounding electrode [250.52(A)(1)] for the separately derived system. **Figure 250–166**

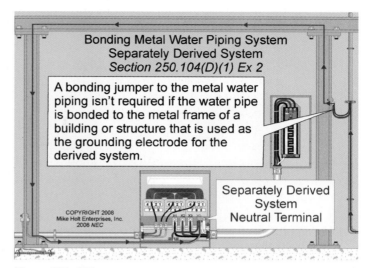

Figure 250–166

(2) Structural Metal. Exposed structural metal interconnected to form the building frame must be bonded to the neutral point of each separately derived system where the grounding electrode conductor is connected.

The bonding jumper must be sized according to Table 250.66, based on the area of the ungrounded secondary conductors.

Exception No. 1: Bonding to the separately derived system isn't required if the metal structural frame serves as the grounding electrode [250.52(A)(2)] for the separately derived system.

250.106 Lightning Protection System. Where a lightning protection system is installed on a building or structure, it must be bonded to the building or structure grounding electrode system. **Figure 250–167**

FPN No. 2: To minimize the likelihood of arcing between metal parts because of induced voltage, metal raceways, enclosures, and other metal parts of electrical equipment may require bonding or spacing from the lightning protection conductors in accordance with NFPA 780, *Standard for the Installation of Lightning Protection Systems.* **Figure 250–168**

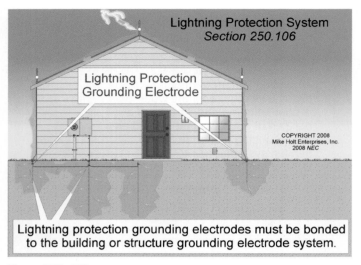

Figure 250–167

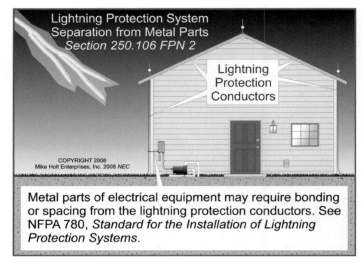

Figure 250–168

Practice Questions

PART V

Use the 2008 *NEC* to answer the following questions.

PART V. BONDING PRACTICE QUESTIONS

1. The non–current-carrying metal parts of service equipment, such as _____, shall be bonded together.

   (a) service raceways or service cable armor
   (b) service equipment enclosures containing service conductors, including meter fittings, boxes, or the like, interposed in the service raceway or armor
   (c) service cable trays
   (d) all of these

2. Bonding jumpers shall be used around _____ knockouts that are punched or otherwise formed so as to impair the electrical connection to ground.

   (a) concentric
   (b) eccentric
   (c) field-punched
   (d) a or b

3. Service equipment, service raceways, and service conductor enclosures shall be bonded _____.

   (a) to the grounded service conductor
   (b) by threaded raceways into enclosures, couplings, hubs, conduit bodies, etc.
   (c) by listed bonding devices with bonding jumpers
   (d) any of these

4. Service raceways threaded into metal service equipment such as bosses (hubs) are considered to be effectively _____ to the service metal enclosure.

   (a) attached
   (b) bonded
   (c) grounded
   (d) none of these

5. Service metal raceways and metal-clad cables are considered effectively bonded when using threadless couplings and connectors that are _____.

   (a) nonmetallic
   (b) made up tight
   (c) sealed
   (d) classified

6. A means external to enclosures for connecting inter-system _____ conductors shall be provided at service equipment and disconnecting means of other buildings or structures.

   (a) bonding
   (b) grounding
   (c) secondary
   (d) a and b

7. When bonding enclosures, metal raceways, frames, and fittings, any nonconductive paint, enamel, or similar coating shall be removed at _____.

   (a) contact surfaces
   (b) threads
   (c) contact points
   (d) all of these

8. For circuits over 250 volts-to-ground, electrical continuity can be maintained between a box or enclosure where no oversized, concentric or eccentric knockouts are encountered, and a metal conduit by _____.

   (a) threadless fittings for cables with metal sheath
   (b) double locknuts on threaded conduit (one inside and one outside the box or enclosure)
   (c) fittings that have shoulders that seat firmly against the box with a locknut on the inside or listed fittings
   (d) all of these

9. Equipment bonding jumpers shall be of copper or other corrosion-resistant material.

   (a) True
   (b) False

10. Equipment bonding jumpers on the supply side of the service shall be no smaller than the sizes shown in _____.

    (a) Table 250.66
    (b) Table 250.122
    (c) Table 310.16
    (d) Table 310.15(B)(6)

11. The equipment bonding jumper on the supply side of services shall be sized according to the _____.

    (a) overcurrent device rating
    (b) service-entrance conductor size
    (c) service-drop size
    (d) load to be served

12. Where service-entrance conductors are paralleled in two or more raceways or cables, the bonding jumper for each raceway or cable shall be based on the size of the _____ in each raceway or cable.

    (a) overcurrent protection for conductors
    (b) grounded conductors
    (c) service-entrance conductors
    (d) sum of all conductors

13. What is the minimum size copper bonding jumper for a service raceway containing 4/0 THHN aluminum conductors?

    (a) 6 AWG aluminum
    (b) 4 AWG aluminum
    (c) 4 AWG copper
    (d) 3 AWG copper

14. A service is supplied by three metal raceways, each containing 600 kcmil ungrounded conductors. Determine the bonding jumper size for each service raceway.

    (a) 1/0 AWG
    (b) 3/0 AWG
    (c) 250 kcmil
    (d) 500 kcmil

15. What is the minimum size copper equipment bonding jumper for a 40A circuit?

    (a) 14 AWG
    (b) 12 AWG
    (c) 10 AWG
    (d) 8 AWG

16. An equipment bonding jumper can be installed on the outside of a raceway, providing the length of the equipment bonding jumper is not more than _____ and the equipment bonding jumper is routed with the raceway.

    (a) 12 in.
    (b) 24 in.
    (c) 36 in.
    (d) 72 in.

17. Metal water piping system(s) shall be bonded to the _____.

    (a) grounded conductor at the service
    (b) service equipment enclosure
    (c) equipment grounding bar or bus at any panelboard within a single occupancy building
    (d) a or b

18. The bonding jumper used to bond the metal water piping system shall be sized in accordance with _____.

    (a) Table 250.66
    (b) Table 250.122
    (c) Table 310.16
    (d) Table 310.15(B)(6)

19. Where isolated metal water piping systems are installed in a multi-occupancy building, the water pipes can be bonded with bonding jumpers sized according to Table 250.122, based on the size of the _____.

    (a) service-entrance conductors
    (b) feeder conductors
    (c) rating of the service equipment overcurrent device
    (d) rating of the overcurrent device supplying the occupancy

20. A building or structure that is supplied by a feeder shall have the interior metal water piping system bonded with a conductor sized in accordance with _____.

    (a) Table 250.66
    (b) Table 250.122
    (c) Table 310.16
    (d) none of these

21. Metal gas piping shall be considered bonded by the equipment grounding conductor of the circuit that is likely to energize the piping.

    (a) True
    (b) False

22. Exposed structural metal interconnected to form a metal building frame that is not intentionally grounded and is likely to become energized, shall be bonded to the _____.

    (a) service equipment enclosure
    (b) grounded conductor at the service
    (c) grounding electrode conductor where of sufficient size
    (d) any of these

23. Lightning protection system ground terminals _____ be bonded to the building grounding electrode system.

    (a) shall
    (b) shall not
    (c) shall be permitted to
    (d) none of these

PART VI. EQUIPMENT GROUNDING AND EQUIPMENT GROUNDING CONDUCTORS

250.110 Fixed Equipment Connected by Permanent Wiring Methods—General.
Exposed metal parts of fixed equipment likely to become energized must be connected to the circuit equipment grounding conductor where the equipment is:

(1) Within 8 ft vertically or 5 ft horizontally of the earth or a grounded metal object

(2) Located in a wet or damp location

(3) In electrical contact with metal

(4) In a hazardous (classified) location [Articles 500 through 517]

(5) Supplied by a wiring method that provides an equipment grounding conductor

(6) Supplied by a 277V or 480V circuit

Exception No. 3: Listed equipment distinctively marked as double-insulated isn't required to be connected to the circuit equipment grounding conductor.

250.112 Fastened in Place or Connected by Permanent Wiring Methods (Fixed).
Except as permitted in 250.112(I), exposed metal parts of equipment and enclosures must be connected to the circuit equipment grounding conductor. Figure 250–169

(I) Remote-Control, Signaling, and Fire Alarm Circuits. Equipment supplied by circuits operating at 50V or less is not required to be connected to the circuit equipment grounding conductor. Figure 250–170

> **Author's Comment:** Class 1 power-limited circuits, Class 2, and Class 3 remote-control and signaling circuits, and fire alarm circuits operating at 50V or less don't need to have any metal parts connected to an equipment grounding conductor.

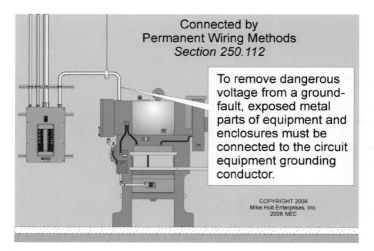

Connected by
Permanent Wiring Methods
Section 250.112

To remove dangerous voltage from a ground-fault, exposed metal parts of equipment and enclosures must be connected to the circuit equipment grounding conductor.

COPYRIGHT 2008
Mike Holt Enterprises, Inc.
2008 *NEC*

Figure 250–169

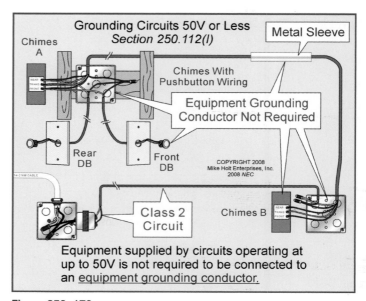

Grounding Circuits 50V or Less
Section 250.112(I)

Chimes A

Metal Sleeve

Chimes With Pushbutton Wiring

Equipment Grounding Conductor Not Required

Rear DB

Front DB

COPYRIGHT 2008
Mike Holt Enterprises, Inc.
2008 *NEC*

Class 2 Circuit

Chimes B

Equipment supplied by circuits operating at up to 50V is not required to be connected to an equipment grounding conductor.

Figure 250–170

250.114 Cord-and-Plug-Connected Equipment. Metal parts of cord-and-plug-connected equipment must be connected to the circuit equipment grounding conductor.

Exception: Listed equipment distinctively marked as double-insulated isn't required to be connected to the circuit equipment grounding conductor.

250.118 Types of Equipment Grounding Conductors. An equipment grounding conductor can be any one or a combination of the following: **Figure 250–171**

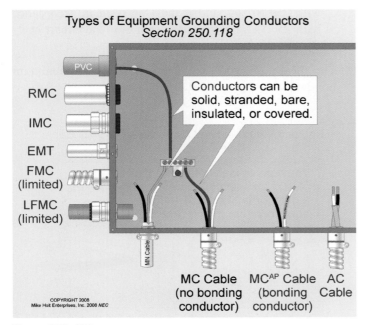

Figure 250–171

FPN: The equipment grounding conductor is intended to serve as the effective ground-fault current path. See 250.2.

Author's Comment: The effective ground-fault path is an intentionally constructed low-impedance conductive path designed to carry fault current from the point of a ground fault on a wiring system to the electrical supply source. Its purpose is to quickly remove dangerous voltage from a ground fault by opening the circuit overcurrent device [250.2]. **Figure 250–172**

(1) A bare or insulated copper or aluminum conductor sized in accordance with 250.122.

Author's Comment: Examples include PVC conduit, Type NM, and Type MC cable with an equipment grounding conductor of the wire type.

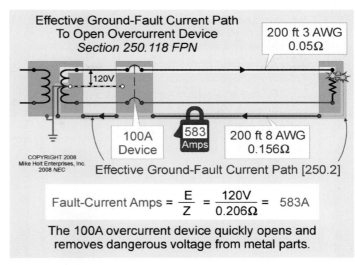

Figure 250–172

(2) Rigid metal conduit (RMC).

(3) Intermediate metal conduit (IMC).

(4) Electrical metallic tubing (EMT).

(5) Listed flexible metal conduit (FMC) where: **Figure 250–173**

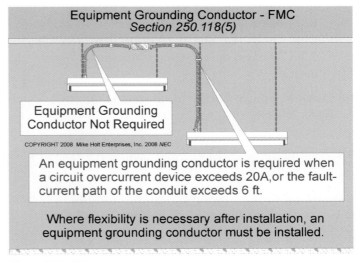

Figure 250–173

a. The conduit terminates in listed fittings.

b. The circuit conductors are protected by an overcurrent device rated 20A or less.

c. The combined length of the flexible conduit doesn't exceed 6 ft. **Figure 250–174**

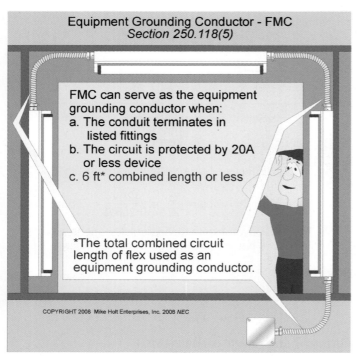

Figure 250–174

d. Where flexibility is required after installation, an equipment grounding conductor of the wire type must be installed with the circuit conductors in accordance with 250.102(E), and it must be sized according to 250.122, based on the rating of the circuit overcurrent device.

(6) Listed liquidtight flexible metal conduit (LFMC) where: **Figure 250–175**

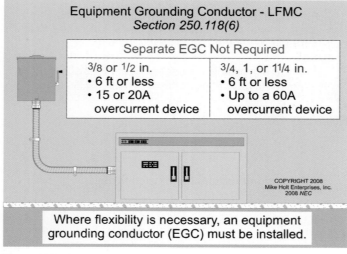

Figure 250–175

a. The conduit terminates in listed fittings.

b. For $\frac{3}{8}$ in. through $\frac{1}{2}$ in., the circuit conductors are protected by an overcurrent device rated 20A or less.

c. For $\frac{3}{4}$ through 1$\frac{1}{4}$ in., the circuit conductors are protected by an overcurrent device rated 60A or less.

d. The combined length of the flexible conduit doesn't exceed 6 ft.

e. Where flexibility is required after installation, an equipment grounding conductor of the wire type must be installed with the circuit conductors in accordance with 250.102(E), and it must be sized in accordance with 250.122, based on the rating of the circuit overcurrent device.

(8) The sheath of Type AC cable containing an aluminum bonding strip. **Figure 250–176**

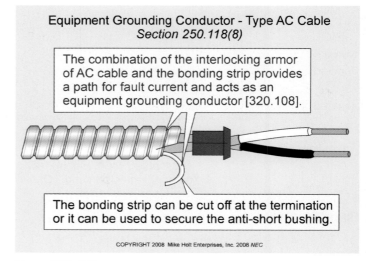

Figure 250–176

Author's Comments:

- The internal aluminum bonding strip isn't an equipment grounding conductor, but it allows the interlocked armor to serve as an equipment grounding conductor because it reduces the impedance of the armored spirals to ensure that a ground fault will be cleared. It's the aluminum bonding strip in combination with the cable armor that creates the circuit equipment grounding conductor. Once the bonding strip exits the cable, it can be cut off because it no longer serves any purpose.

- The effective ground-fault current path must be maintained by the use of fittings specifically listed for Type AC cable [320.40]. See 300.12, 300.15, and 320.100.

(9) The copper sheath of Type MI cable.

(10) The sheath of Type MC cable where listed and identified for grounding:

   a. Interlocked Type MC cable containing a bare aluminum bonding strip in direct contact with the interlocked metal armor that is listed and identified for grounding. **Figure 250–177**

   b. Smooth or corrugated-tube Type MC cable without an equipment grounding conductor.

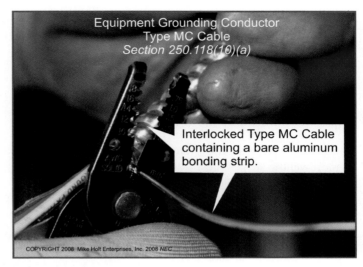

Equipment Grounding Conductor
Type MC Cable
Section 250.118(10)(a)

Interlocked Type MC Cable containing a bare aluminum bonding strip.

COPYRIGHT 2008 Mike Holt Enterprises, Inc. 2008 NEC

Figure 250–178

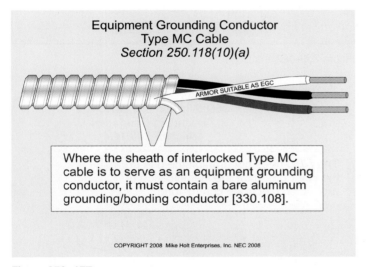

Equipment Grounding Conductor
Type MC Cable
Section 250.118(10)(a)

ARMOR SUITABLE AS EGC

Where the sheath of interlocked Type MC cable is to serve as an equipment grounding conductor, it must contain a bare aluminum grounding/bonding conductor [330.108].

COPYRIGHT 2008 Mike Holt Enterprises, Inc. NEC 2008

Figure 250–177

Author's Comment: MC^{AP®} cable is Type MC cable constructed with THHN copper insulated circuit conductors and interlocked armor that is listed and identified for grounding. After considering potential cable applications, Southwire decided to name the product MC^{AP} cable to reflect the all-purpose aspect of the product. The "AP" in MC^{AP} stands for All-Purpose, which means that MC^{AP} cable can be used in both MC and AC cable applications. The dual use is permitted because the armor of MC^{AP} cable is listed and identified as a suitable equipment grounding path, unlike conventional MC cable.

An aluminum grounding/bonding conductor in direct contact with the interlocked armor throughout the entire cable length allows the armor of MC^{AP} cable to serve as an equipment grounding conductor.

Once the bare aluminum grounding/bonding conductor exits the cable, it can be cut off because it no longer serves any purpose. The effective ground-fault current path must be maintained by the use of fittings specifically listed for Type MC^{AP} cable [330.40]. See 300.12, 300.15, and 330.100. **Figure 250–178**

(11) Metallic cable trays where continuous maintenance and supervision ensure only qualified persons will service the cable tray [392.3(C)], with cable tray and fittings identified for grounding and the cable tray, fittings, and raceways are bonded using bolted mechanical connectors or bonding jumpers sized and installed in accordance with 250.102 [392.7]. **Figure 250–179**

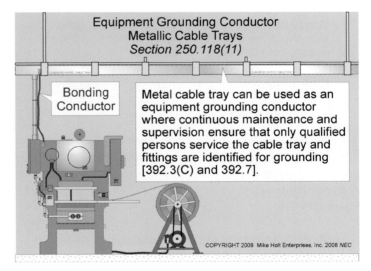

Equipment Grounding Conductor
Metallic Cable Trays
Section 250.118(11)

Bonding Conductor

Metal cable tray can be used as an equipment grounding conductor where continuous maintenance and supervision ensure that only qualified persons service the cable tray and fittings are identified for grounding [392.3(C) and 392.7].

COPYRIGHT 2008 Mike Holt Enterprises, Inc. 2008 NEC

Figure 250–179

(13) Listed electrically continuous metal raceways, such as metal wireways [Article 376] or strut-type channel raceways [384.60].

(14) Surface metal raceways listed for grounding [Article 386].

250.119 Identification of Equipment Grounding Conductors.

Unless required to be insulated, equipment grounding conductors can be bare, covered or insulated. Insulated equipment grounding conductors must have a continuous outer finish that is either green or green with one or more yellow stripes.

Conductors with insulation that is green, or green with one or more yellow stripes must not be used for an ungrounded or neutral conductor. **Figure 250–180**

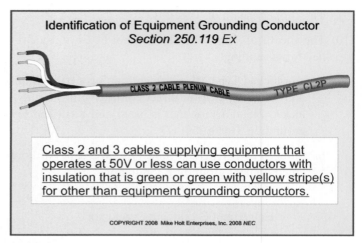

Identification of Equipment Grounding Conductor
Section 250.119 Ex

Class 2 and 3 cables supplying equipment that operates at 50V or less can use conductors with insulation that is green or green with yellow stripe(s) for other than equipment grounding conductors.

COPYRIGHT 2008 Mike Holt Enterprises, Inc. 2008 NEC

Figure 250–181

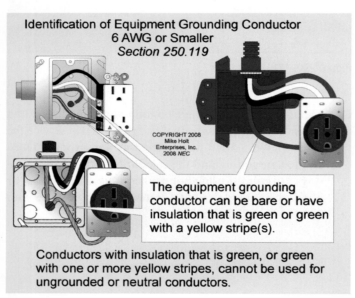

Identification of Equipment Grounding Conductor
6 AWG or Smaller
Section 250.119

The equipment grounding conductor can be bare or have insulation that is green or green with a yellow stripe(s).

Conductors with insulation that is green, or green with one or more yellow stripes, cannot be used for ungrounded or neutral conductors.

Figure 250–180

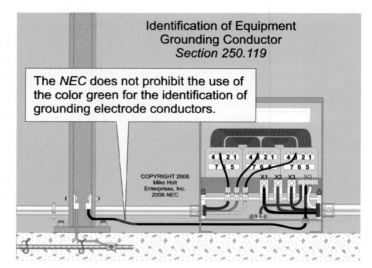

Identification of Equipment Grounding Conductor
Section 250.119

The NEC does not prohibit the use of the color green for the identification of grounding electrode conductors.

COPYRIGHT 2008 Mike Holt Enterprises, Inc. 2008 NEC

Figure 250–182

Exception: Class 2 or Class 3 cables supplying equipment that operates at less than 50V can use conductors with insulated green or green with one or more yellow stripes for other than equipment grounding conductors [250.20(A) and 250.112(I)]. **Figure 250–181**

> **Author's Comment:** The NEC neither requires nor prohibits the use of the color green for the identification of grounding electrode conductors. **Figure 250–182**

(A) Conductors Larger Than 6 AWG.

(1) Identified Where Accessible. Insulated equipment grounding conductors larger than 6 AWG can be permanently reidentified at the time of installation at every point where the conductor is accessible.

Exception: Identification of equipment grounding conductors larger than 6 AWG in conduit bodies is not required.

(2) Identification Method. Equipment grounding conductor identification must encircle the conductor by: **Figure 250–183**

   a. Removing insulation at termination

   b. Coloring the insulation green at termination

   c. Marking the insulation at termination with green tape or green adhesive labels

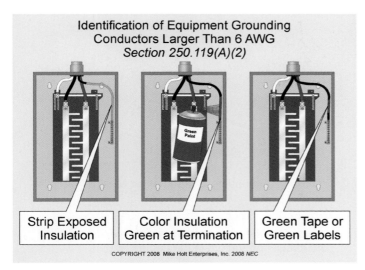

Figure 250–183

250.122 Sizing Equipment Grounding Conductor.

(A) General. Equipment grounding conductors of the wire type must be sized not smaller than shown in Table 250.122 based on the rating of the circuit overcurrent device; however the circuit equipment grounding conductor is not required to be larger than the circuit conductors. **Figure 250–184**

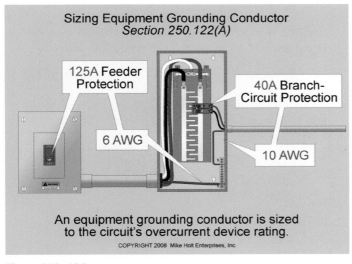

Figure 250–184

Table 250.122 Sizing Equipment Grounding Conductor	
Overcurrent Device Rating	Copper Conductor
15A	14 AWG
20A	12 AWG
30A—60A	10 AWG
70A—100A	8 AWG
110A—200A	6 AWG
225A—300A	4 AWG
350A—400A	3 AWG
450A—500A	2 AWG
600A	1 AWG
700A—800A	1/0 AWG
1,000A	2/0 AWG
1,200A	3/0 AWG

(B) Increased in Size. When ungrounded circuit conductors are increased in size, the circuit equipment grounding conductor must be proportionately increased in size according to the circular mil area of the ungrounded conductors.

> **Author's Comment:** Ungrounded conductors are sometimes increased in size to accommodate for conductor voltage drop, harmonic current heating, short-circuit rating, or simply for future capacity.

> **Question:** If the ungrounded conductors for a 40A circuit are increased in size from 8 AWG to 6 AWG, the circuit equipment grounding conductor must be increased in size from 10 AWG to _____. **Figure 250–185**
>
> (a) 10 AWG   (b) 8 AWG   (c) 6 AWG   (d) 4 AWG
>
> **Answer:** (b) 8 AWG
>
> The circular mil area of 6 AWG is 59 percent more than 8 AWG (26,240 cmil/16,510 cmil) [Chapter 9, Table 8].
>
> According to Table 250.122, the circuit equipment grounding conductor for a 40A overcurrent device will be 10 AWG (10,380 cmil), but the circuit equipment grounding conductor for this circuit must be increased in size by a multiplier of 1.59.
>
> Conductor Size = 10,380 cmil x 1.59
> Conductor Size = 16,504 cmil
> Conductor Size = 8 AWG, Chapter 9, Table 8

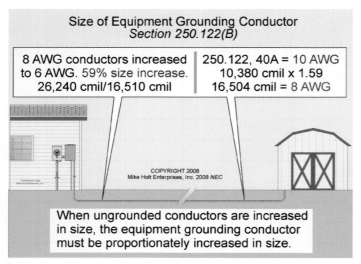

Figure 250–185

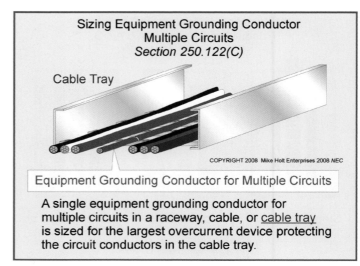

Figure 250–187

(C) Multiple Circuits. When multiple circuits are installed in the same raceway, cable, or cable tray, only one equipment grounding conductor is required for the multiple circuits, sized in accordance with 250.122, based on the rating of the largest circuit overcurrent device. **Figures 250–186** and **250–187**

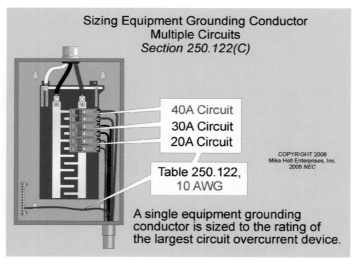

Figure 250–186

Author's Comment: Single conductors used as equipment grounding conductors in cable trays must be sized 4 AWG or larger [392.3(B)(1)(c)].

(D) Motor Branch Circuits.

(1) General. The equipment grounding conductor, where of the wire type, must be sized in accordance with Table 250.122, based on the rating of the motor circuit branch-circuit short-circuit and ground-fault overcurrent device, but this conductor is not required to be larger than the circuit conductors [250.122(A)].

Question: What size equipment grounding conductor is required for a 2 hp, 230V, single-phase motor? **Figure 250–188**

(a) 14 AWG   (b) 12 AWG   (c) 10 AWG   (d) 8 AWG

Answer: (a) 14 AWG

Step 1: Determine the branch-circuit conductor [Table 310.16 and 430.22(A)

2 hp, 230V Motor FLC = 12A [Table 430.248]

12A x 1.25 = 15A, 14 AWG, rated 20A at 75°C [Table 310.16]

Step 2: Determine the branch-circuit protection [240.6(A), 430.52(C)(1), and Table 430.248]

12A x 2.50 = 30A

Step 3: The circuit equipment grounding conductor must be sized to the 30A overcurrent device—10 AWG [Table 250.122], but it's not required to be sized larger than the circuit conductors —14 AWG.

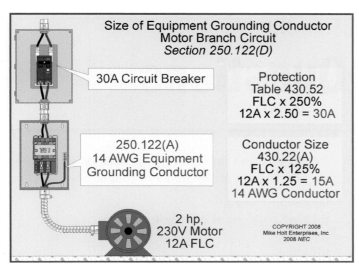

Figure 250–188

(F) Parallel Runs. When circuit conductors are run in parallel [310.4], an equipment grounding conductor must be installed with each parallel conductor set and it must be sized in accordance with Table 250.122, based on the rating of the circuit overcurrent device, but this conductor is not required to be larger than the circuit conductors [250.112(A)]. **Figure 250–189**

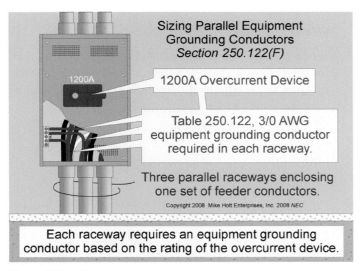

Figure 250–189

(G) Feeder Tap Conductors. Equipment grounding conductors for feeder taps must be sized in accordance with Table 250.122, based on the ampere rating of the overcurrent device ahead of the feeder, but in no case is it required to be larger than the feeder tap conductors. **Figure 250–190**

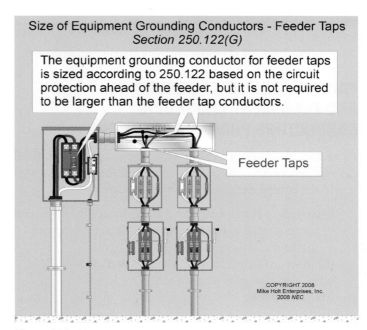

Figure 250–190

PART VI | Practice Questions

Use the 2008 *NEC* to answer the following questions.

PART VI. EQUIPMENT GROUNDING AND EQUIPMENT GROUNDING CONDUCTORS PRACTICE QUESTIONS

1. Exposed non–current-carrying metal parts of fixed equipment likely to become energized shall be connected to the equipment conductor where located _____.

   (a) within 8 ft vertically or 5 ft horizontally of ground or grounded metal objects and subject to contact by persons
   (b) in wet or damp locations and not isolated
   (c) in electrical contact with metal
   (d) any of these

2. Listed FMC can be used as the equipment grounding conductor if the length in any ground return path does not exceed 6 ft and the circuit conductors contained in the conduit are protected by overcurrent devices rated at _____ or less.

   (a) 15A
   (b) 20A
   (c) 30A
   (d) 60A

3. Listed FMC and LFMC shall contain an equipment grounding conductor if the raceway is installed for the reason of _____.

   (a) physical protection
   (b) flexibility after installation
   (c) protection from moisture
   (d) communications systems

4. An equipment grounding conductor shall be identified by _____.

   (a) a continuous outer finish that is green
   (b) being bare
   (c) a continuous outer finish that is green with one or more yellow stripes
   (d) any of these

5. Conductors with the color _____ insulation shall not be used for ungrounded or grounded conductors.

   (a) green
   (b) green with one or more yellow stripes
   (c) a or b
   (d) white

6. The equipment grounding conductor shall not be required to be larger than the circuit conductors.

   (a) True
   (b) False

7. When ungrounded circuit conductors are increased in size, the equipment grounding conductor must be proportionately increased in size according to the _____ of the ungrounded conductors.

   (a) ampacity
   (b) circular mil area
   (c) diameter
   (d) none of these

8. When a single equipment grounding conductor is used for multiple circuits in the same raceway, cable or cable tray, the single equipment grounding conductor shall be sized according to the _____.

   (a) combined rating of all the overcurrent devices
   (b) largest overcurrent device of the multiple circuits
   (c) combined rating of all the loads
   (d) any of these

9. Equipment grounding conductors for motor branch circuits shall be sized in accordance with Table 250.122, based on the rating of the _____ device.

   (a) motor overload
   (b) motor over-temperature
   (c) motor short-circuit and ground-fault protective
   (d) feeder overcurrent protection

10. Where conductors are run in parallel, the equipment grounding conductor must be installed with each parallel conductor set and it's not required to be larger than the circuit conductors.

    (a) True
    (b) False

11. Equipment grounding conductors for feeder taps are not required to be larger than the tap conductors.

    (a) True
    (b) False

12. The terminal of a wiring device for the connection of the equipment grounding conductor shall be identified by a green-colored, _____.

    (a) not readily removable terminal screw with a hexagonal head
    (b) hexagonal, not readily removable terminal nut
    (c) pressure wire connector
    (d) any of these

ARTICLE 250

Grounding and Bonding— Part VII. Methods of Equipment Grounding

PART VII. METHODS OF EQUIPMENT GROUNDING

250.130 Replacing Nongrounding Receptacles.

(C) Nongrounding Receptacle Replacement. Where a nongrounding receptacle is replaced with a grounding-type receptacle from an outlet box that doesn't contain an equipment grounding conductor, the grounding contacts of the receptacle must be connected to one of the following: **Figure 250–191**

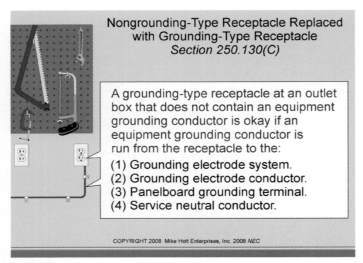

Nongrounding-Type Receptacle Replaced with Grounding-Type Receptacle
Section 250.130(C)

A grounding-type receptacle at an outlet box that does not contain an equipment grounding conductor is okay if an equipment grounding conductor is run from the receptacle to the:

(1) Grounding electrode system.
(2) Grounding electrode conductor.
(3) Panelboard grounding terminal.
(4) Service neutral conductor.

COPYRIGHT 2008 Mike Holt Enterprises, Inc. 2008 NEC

Figure 250–191

(1) Grounding electrode system [250.50]

(2) Grounding electrode conductor

(3) Panelboard equipment grounding terminal

(4) Service neutral conductor

> **FPN:** A grounding-type receptacle can replace a nongrounding type receptacle, without having the grounding terminal connected to an equipment grounding conductor, if the receptacle is GFCI-protected and marked in accordance with 406.3(D)(3). **Figure 250–192**

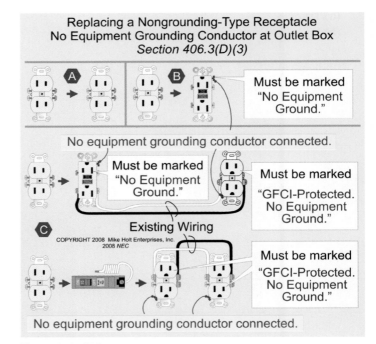

Replacing a Nongrounding-Type Receptacle
No Equipment Grounding Conductor at Outlet Box
Section 406.3(D)(3)

Must be marked "No Equipment Ground."

No equipment grounding conductor connected.

Must be marked "No Equipment Ground."

Must be marked "GFCI-Protected. No Equipment Ground."

Existing Wiring

COPYRIGHT 2008 Mike Holt Enterprises, Inc.
2008 NEC

Must be marked "GFCI-Protected. No Equipment Ground."

No equipment grounding conductor connected.

Figure 250–192

250.134 Equipment Fastened in Place or Connected by Wiring Methods.
Unless connected to the neutral conductor at services or separately derived systems as permitted or required by 250.142, metal parts of equipment, raceways, and enclosures must be connected to an equipment grounding conductor by one of the following methods:

(A) Equipment Grounding Conductor Types. By connecting to equipment grounding conductors identified in 250.118.

(B) With Circuit Conductors. Where an equipment grounding conductor of the wire type is installed, it must be installed in the same raceway, cable tray, trench, cable, or cord with the circuit conductors in accordance with 300.3(B), except as permitted by 250.102(E). **Figures 250–193** and **250–194**

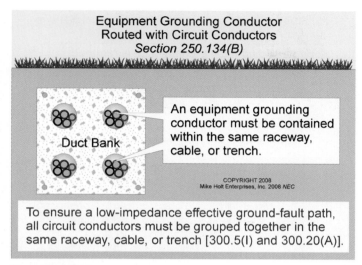

Figure 250–193

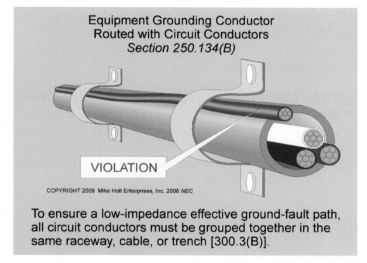

Figure 250–194

250.136 Equipment Considered Grounded.

(A) Equipment Secured to Grounded Metal Supports. The structural metal frame of a building must not be used as the required equipment grounding conductor.

250.138 Cord-and-Plug-Connected Equipment.

(A) Equipment Grounding Conductor. Metal parts of cord-and-plug-connected equipment must be connected to an equipment grounding conductor that terminates to a grounding-type attachment plug.

250.140 Ranges, Ovens, and Clothes Dryers. The frames of electric ranges, wall-mounted ovens, counter-mounted cooking units, clothes dryers, and outlet boxes that are part of the circuit for these appliances must be connected to the equipment grounding conductor [250.134(A)]. Figure 250–195

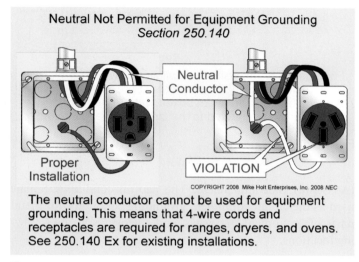

Figure 250–195

CAUTION: *Ranges, dryers, and ovens have their metal cases connected to the neutral conductor at the factory. This neutral-to-case connection must be removed when these appliances are installed in new construction, and a 4-wire cord and receptacle must be used [250.142(B)].*

Exception: For existing installations where an equipment grounding conductor isn't present in the outlet box, the frames of electric ranges, wall-mounted ovens, counter-mounted cooking units, clothes dryers, and outlet boxes that are part of the circuit for these appliances may be connected to the neutral conductor. Figure 250–196

250.142 Use of Neutral Conductor for Equipment Grounding.

Author's Comment: To remove dangerous voltage on metal parts from a ground fault, the metal parts of electrical raceways, cables, enclosures, and equipment must be connected to an equipment grounding conductor of a type recognized in 250.118 in accordance with 250.4(A)(3).

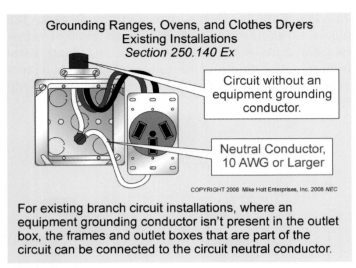

Figure 250–196

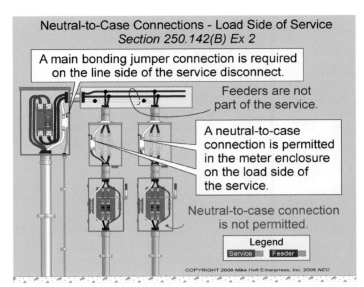

Figure 250–197

(A) Supply Side Equipment. The neutral conductor can be used as the circuit equipment grounding conductor for metal parts of equipment, raceways, and enclosures at the following locations:

(1) Service Equipment. On the supply side or within the enclosure of the service disconnect in accordance with 250.24(B).

(3) Separately Derived Systems. At the source of a separately derived system or within the enclosure of the system disconnecting means in accordance with 250.30(A)(1).

> **DANGER:** *Failure to install the system bonding jumper as required by 250.30(A)(1) creates a condition where dangerous touch voltage from a ground fault will not be removed.*

(B) Load Side Equipment. Except for service equipment and separately derived systems, the neutral conductor must not serve as an equipment grounding conductor on the load side of service equipment.

Exception No. 1: In existing installations, the frames of ranges, wall-mounted ovens, counter-mounted cooking units, and clothes dryers can be connected to the neutral conductor according to 250.140 Ex.

Exception No. 2: The neutral conductor can be connected to meter socket enclosures on the load side of the service disconnecting means if: **Figure 250–197**

(1) Ground-fault protection is not provided on service equipment,

(2) Meter socket enclosures are immediately adjacent to the service disconnecting means, and

(3) The neutral conductor is sized in accordance with 250.122, based on the ampere rating of the occupancy's feeder overcurrent device.

250.146 Connecting Receptacle Grounding Terminal to Metal Enclosure. An equipment bonding jumper sized in accordance with 250.122, based on the rating of the circuit overcurrent device, must connect the grounding terminal of a receptacle to a metal box, except as permitted for (A) through (D). **Figure 250–198**

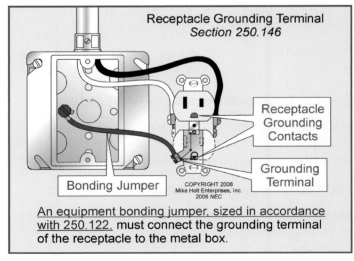

Figure 250–198

Author's Comment: The *NEC* does not restrict the position of the receptacle grounding terminal; it can be up, down, or sideways. *Code* proposals to specify the mounting position of receptacles have always been rejected. **Figure 250–199**

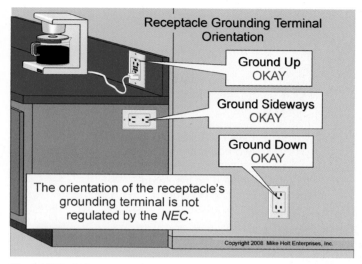

Figure 250–199

(A) Surface-Mounted Box. An equipment bonding jumper from a receptacle to a metal box that is surface mounted is not required where there is direct metal-to-metal contact between the device yoke and the metal box. To ensure a suitable bonding path between the device yoke and a metal box, at least one of the insulating retaining washers on the yoke screw must be removed. **Figure 250–200**

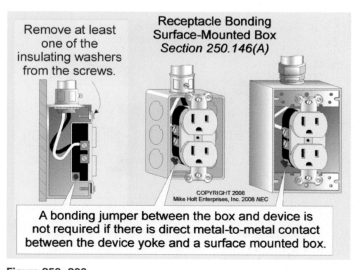

Figure 250–200

An equipment bonding jumper is not required for receptacles attached to listed exposed work covers when the receptacle is attached to the cover with at least two fasteners that have a thread locking or screw locking means, and the cover mounting holes are located on a flat non-raised portion of the cover. **Figure 250–201**

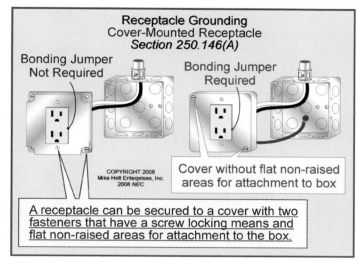

Figure 250–201

(B) Self-Grounding Receptacles. Receptacle yokes listed as self-grounding are designed to establish the bonding path between the device yoke and a metal box via the two metal mounting screws. **Figure 250–202**

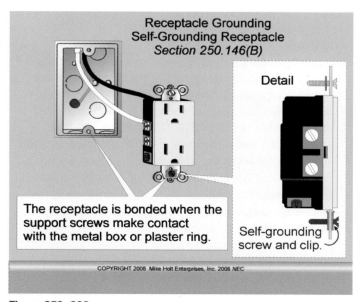

Figure 250–202

(C) Floor Boxes. Listed floor boxes are designed to establish the bonding path between the device yoke and a metal box.

(D) Isolated Ground Receptacles. Where installed for the reduction of electrical noise, the grounding terminal of an isolated ground receptacle must be connected to an insulated equipment grounding conductor run with the circuit conductors. **Figure 250–203**

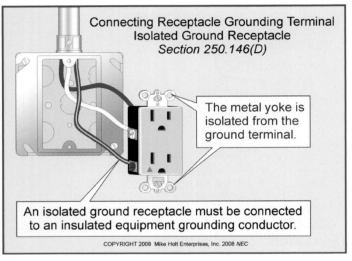

Figure 250–203

The circuit equipment grounding conductor is permitted to pass through panelboards [408.40 Ex], boxes, wireways, or other enclosures [250.148 Ex] without a connection to the enclosure as long as it terminates at an equipment grounding conductor terminal of the derived system or service.

CAUTION:
> *Type AC Cable—Type AC cable containing an insulated equipment grounding conductor of the wire type can be used to supply receptacles having insulated grounding terminals because the metal armor of the cable is listed as an equipment grounding conductor [250.118(8)].* **Figure 250–204**
>
> *Type MC Cable—The metal armor sheath of interlocked Type MC cable containing an insulated equipment grounding conductor isn't listed as an equipment grounding conductor. Therefore, this wiring method with a single equipment grounding conductor can't supply an isolated ground receptacle installed in a metal box (because the box is not connected to an equipment grounding conductor). However, Type MC cable with two insulated equipment grounding conductors is acceptable, since one equipment grounding conductor connects to the metal box and the other to the isolated ground receptacle.* **Figure 250–204**

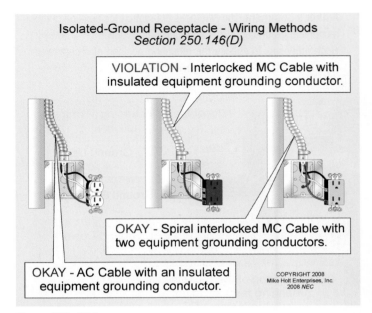

Figure 250–204

> *The armor assembly of interlocked Type MC^{AP} cable with a 10 AWG bare aluminum grounding/bonding conductor running just below the metal armor is listed to serve as an equipment grounding conductor in accordance with 250.118(10) (a).*
>
> *Nonmetallic Boxes—Because the grounding terminal of an isolated ground receptacle is insulated from the metal mounting yoke, a metal faceplate must not be used when an isolated ground receptacle is installed in a nonmetallic box. The reason is that the metal faceplate is not connected to an equipment grounding conductor [406.2(D)(2)].* **Figure 250–205**

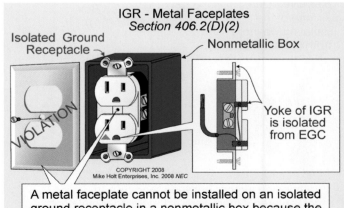

Figure 250–205

Author's Comment: When should an isolated ground receptacle be installed and how should the isolated-ground system be designed? These questions are design issues and must not be answered based on the *NEC* alone [90.1(C)]. In most cases, using isolated ground receptacles is a waste of money. For example, IEEE 1100, *Powering and Grounding Electronic Equipment* (Emerald Book) states: "The results from the use of the isolated-ground method range from no observable effects, the desired effects, or worse noise conditions than when standard equipment bonding configurations are used to serve electronic load equipment [8.5.3.2]."

In reality, few electrical installations truly require an isolated-ground system. For those systems that can benefit from an isolated-ground system, engineering opinions differ as to what is a proper design. Making matters worse—of those properly designed, few are correctly installed and even fewer are properly maintained. For more information on how to properly ground electronic equipment, go to: www.mikeholt.com, click on the "Technical" link, and then visit the "Power Quality" page.

250.148 Continuity and Attachment of Equipment Grounding Conductors in Boxes.

Where circuit conductors are spliced or terminated on equipment within a metal box, the equipment grounding conductor associated with those circuits must be connected to the box in accordance with the following: **Figure 250–206**

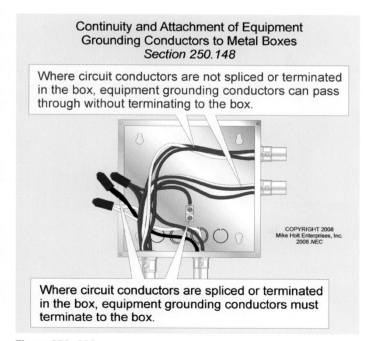

Continuity and Attachment of Equipment
Grounding Conductors to Metal Boxes
Section 250.148

Where circuit conductors are not spliced or terminated in the box, equipment grounding conductors can pass through without terminating to the box.

COPYRIGHT 2008 Mike Holt Enterprises, Inc. 2008 NEC

Where circuit conductors are spliced or terminated in the box, equipment grounding conductors must terminate to the box.

Figure 250–206

Exception: The circuit equipment grounding conductor for an isolated ground receptacle installed in accordance with 250.146(D) isn't required to terminate to a metal box. **Figure 250–207**

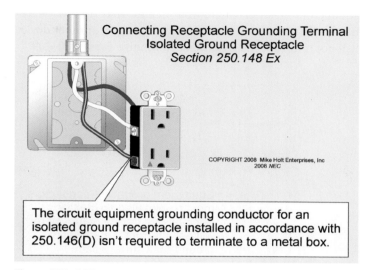

Connecting Receptacle Grounding Terminal
Isolated Ground Receptacle
Section 250.148 Ex

COPYRIGHT 2008 Mike Holt Enterprises, Inc
2008 NEC

The circuit equipment grounding conductor for an isolated ground receptacle installed in accordance with 250.146(D) isn't required to terminate to a metal box.

Figure 250–207

(A) Splicing. Equipment grounding conductors must be spliced together with a device listed for the purpose [110.14(B)]. **Figure 250–208**

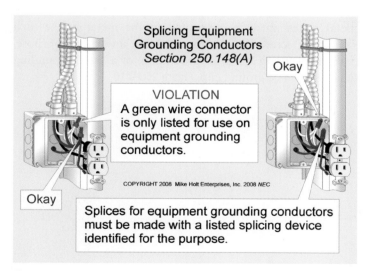

Splicing Equipment
Grounding Conductors
Section 250.148(A)

Okay

VIOLATION
A green wire connector is only listed for use on equipment grounding conductors.

COPYRIGHT 2008 Mike Holt Enterprises, Inc. 2008 NEC

Okay

Splices for equipment grounding conductors must be made with a listed splicing device identified for the purpose.

Figure 250–208

Author's Comment: Wire connectors of any color can be used with equipment grounding conductor splices, but green wire connectors can only be used with equipment grounding conductors.

(B) Equipment Grounding Continuity. Equipment grounding conductors must terminate in a manner that the disconnection or the removal of a receptacle, luminaire, or other device will not interrupt the grounding continuity. **Figure 250–209**

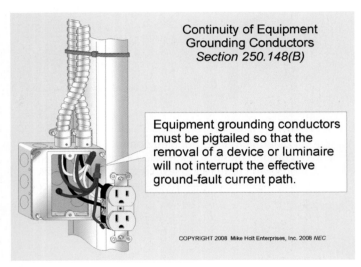

Continuity of Equipment
Grounding Conductors
Section 250.148(B)

Equipment grounding conductors must be pigtailed so that the removal of a device or luminaire will not interrupt the effective ground-fault current path.

COPYRIGHT 2008 Mike Holt Enterprises, Inc. 2008 *NEC*

Figure 250–209

(C) Metal Boxes. Equipment grounding conductors within metal boxes must be connected to the metal box with a grounding screw that is not used for any other purpose, equipment fitting listed for grounding, or a listed grounding device such as a ground clip. **Figure 250–210**

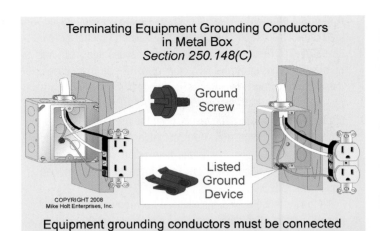

Terminating Equipment Grounding Conductors
in Metal Box
Section 250.148(C)

Ground Screw

Listed Ground Device

COPYRIGHT 2008
Mike Holt Enterprises, Inc.

Equipment grounding conductors must be connected to a metal box by a grounding screw that is not used for any other purpose or a listed grounding device.

Figure 250–210

Author's Comment: Equipment grounding conductors aren't permitted to terminate to a screw that secures a plaster ring. **Figure 250–211**

Terminating Equipment Grounding
Conductors in a Metal Box
Section 250.148(C)

VIOLATION

Equipment grounding conductors must be connected to a metal box by a screw or fitting that is not used for any other purpose.

COPYRIGHT 2008
Mike Holt Enterprises, Inc. 2008 *NEC*

Figure 250–211

PART VII Practice Questions

Use the 2008 *NEC* to answer the following questions.

PART VII. METHODS OF EQUIPMENT GROUNDING PRACTICE QUESTIONS

1. The structural metal frame of a building can be used as the required circuit equipment grounding conductor.

   (a) True
   (b) False

2. Metal parts of cord-and-plug-connected equipment shall be connected to an equipment grounding conductor that terminates to a grounding-type attachment plug.

   (a) True
   (b) False

3. A grounded circuit conductor is permitted to ground non–current-carrying metal parts of equipment, raceways, and other enclosures on the supply side or within the enclosure of the ac service-disconnecting means.

   (a) True
   (b) False

4. It shall be permissible to ground meter enclosures located near the service disconnecting means to the _____ circuit conductor on the load side of the service disconnect, if service ground-fault protection is not provided.

   (a) grounding
   (b) bonding
   (c) grounded
   (d) phase

5. A(n) _____ shall be used to connect the grounding terminal of a grounding-type receptacle to a grounded box.

   (a) equipment bonding jumper
   (b) grounded conductor jumper
   (c) a or b
   (d) a and b

6. Where the box is mounted on the surface, direct metal-to-metal contact between the device yoke and the box shall be permitted to ground the receptacle to the box if at least _____ of the insulating washers of the receptacle is (are) removed.

   (a) one
   (b) two
   (c) three
   (d) none of these

7. A listed exposed work cover can be the grounding and bonding means when the device is attached to the cover with at least _____ fastener(s) and the cover mounting holes are located on a non-raised portion of the cover.

   (a) one
   (b) two
   (c) three
   (d) none of these

8. Receptacle yokes designed and _____ as self-grounding can establish the grounding circuit between the device yoke and a grounded outlet box.

   (a) approved
   (b) advertised
   (c) listed
   (d) installed

9. The receptacle grounding terminal of an isolated ground receptacle shall be connected to a(n) _____ equipment grounding conductor run with the circuit conductors.

   (a) insulated
   (b) covered
   (c) bare
   (d) solid

10. Where circuit conductors are spliced or terminated on equipment within a box, any equipment grounding conductors associated with those circuit conductors shall be connected to the box with devices suitable for the use.

    (a) True
    (b) False

11. The arrangement of grounding connections shall be such that the disconnection or the removal of a receptacle, luminaire, or other device does not interrupt the grounding continuity.

    (a) True
    (b) False

12. A connection between equipment grounding conductors and a metal box shall be by _____.

    (a) a grounding screw used for no other purpose
    (b) equipment listed for grounding
    (c) a listed grounding device
    (d) any of these

300.10 Electrical Continuity. Metal raceways, cables, boxes, fittings, cabinets, and enclosures for conductors must be metallically joined together to form a continuous low-impedance fault current path capable of carrying any fault current likely to be imposed on it [110.10, 250.4(A)(3), and 250.122]. **Figure 300–1**

Metal raceways and cable assemblies must be mechanically secured to boxes, fittings, cabinets, and other enclosures.

Exception No. 1: Short lengths of metal raceways used for the support or protection of cables aren't required to be electrically continuous, nor are they required to be connected to an equipment grounding conductor of a type recognized in 250.118 [250.86 Ex 2 and 300.12 Ex]. **Figure 300–2**

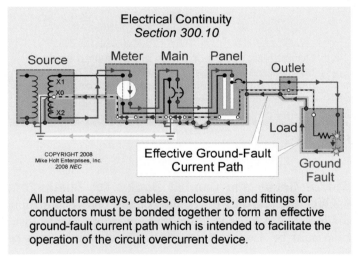

All metal raceways, cables, enclosures, and fittings for conductors must be bonded together to form an effective ground-fault current path which is intended to facilitate the operation of the circuit overcurrent device.

Figure 300–1

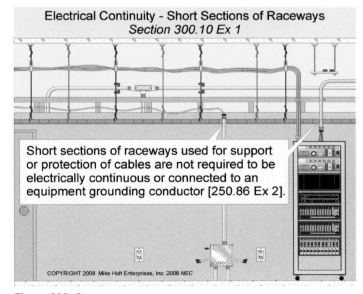

Figure 300–2

ARTICLE 314

Outlet, Device, Pull and Junction Boxes, Conduit Bodies, and Handhole Enclosures

314.4 Metal Boxes. Metal boxes containing circuits that operate at 50V or more must be connected to an equipment grounding conductor of a type listed in 250.118 [250.112(I)]. **Figure 314–1**

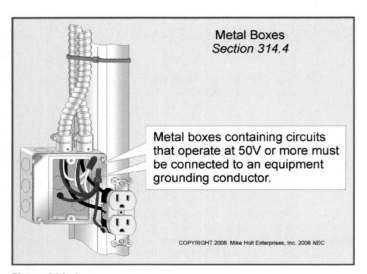

Metal Boxes
Section 314.4

Metal boxes containing circuits that operate at 50V or more must be connected to an equipment grounding conductor.

COPYRIGHT 2008 Mike Holt Enterprises, Inc. 2008 NEC

Figure 314–1

314.25 Covers and Canopies. When the installation is complete, each outlet box must be provided with a cover or faceplate, unless covered by a fixture canopy, lampholder, or similar device. **Figure 314–2**

(A) Nonmetallic or Metallic. Nonmetallic covers are permitted on any box, but metal covers are only permitted where they can be connected to an equipment grounding conductor of a type recognized in 250.118 in accordance with 250.110 [250.4(A)(3)].

> **Author's Comment:** Metal switch faceplates [404.9(B)] and metal receptacle faceplates [406.5(A)] must be connected to an equipment grounding conductor.

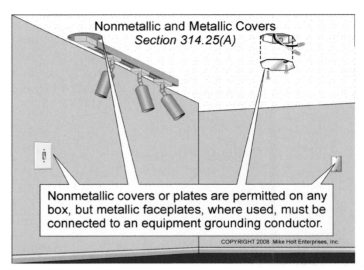

Nonmetallic and Metallic Covers
Section 314.25(A)

Nonmetallic covers or plates are permitted on any box, but metallic faceplates, where used, must be connected to an equipment grounding conductor.

COPYRIGHT 2008 Mike Holt Enterprises, Inc.

Figure 314–2

314.28 Boxes and Conduit Bodies for Conductors 4 AWG and Larger. Boxes and conduit bodies containing conductors 4 AWG and larger that are required to be insulated must be sized so the conductor insulation will not be damaged.

(C) Covers. Pull boxes, junction boxes, and conduit bodies must have a cover suitable for the conditions. Nonmetallic covers are permitted on any box, but metal covers are only permitted where they can be connected to an equipment grounding conductor of a type recognized in 250.118 in accordance with 250.110 [250.4(A)(3)]. **Figure 314–3**

314.30 Handhole Enclosures. Handhole enclosures must be identified for underground use, and be designed and installed to withstand all loads likely to be imposed.

(D) Covers. Metal covers and other exposed conductive surfaces of handhole enclosures must be connected to an equipment grounding conductor sized to the overcurrent device in accordance with 250.122. **Figure 314–4**

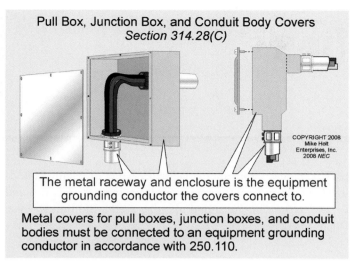

Pull Box, Junction Box, and Conduit Body Covers
Section 314.28(C)

COPYRIGHT 2008
Mike Holt
Enterprises, Inc.
2008 NEC

The metal raceway and enclosure is the equipment grounding conductor the covers connect to.

Metal covers for pull boxes, junction boxes, and conduit bodies must be connected to an equipment grounding conductor in accordance with 250.110.

Figure 314–3

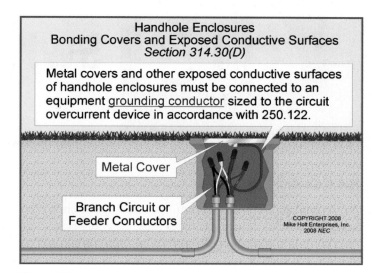

Handhole Enclosures
Bonding Covers and Exposed Conductive Surfaces
Section 314.30(D)

Metal covers and other exposed conductive surfaces of handhole enclosures must be connected to an equipment grounding conductor sized to the circuit overcurrent device in accordance with 250.122.

Metal Cover

Branch Circuit or Feeder Conductors

COPYRIGHT 2008
Mike Holt Enterprises, Inc.
2008 NEC

Figure 314–4

404.9 Switch Faceplates.

(B) Grounding. The metal mounting yokes for switches, dimmers, and similar control switches must be connected to an equipment grounding conductor of a type recognized in 250.118, whether or not a metal faceplate is installed. The metal mounting yoke is considered part of the effective ground-fault current path by one of the following means:

(1) Mounting Screw. The switch is mounted with metal screws to a metal box or a metal cover connected to an equipment grounding conductor of a type recognized in 250.118. **Figure 404–1**

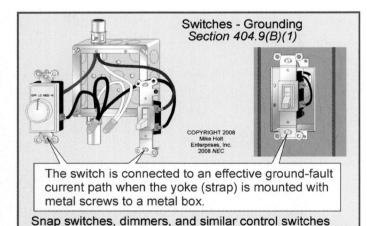

The switch is connected to an effective ground-fault current path when the yoke (strap) is mounted with metal screws to a metal box.

Snap switches, dimmers, and similar control switches must be connected to an equipment grounding conductor whether or not a metal faceplate is installed [404.9(B)].

Figure 404–1

Author's Comment: Direct metal-to-metal contact between the device yoke of a switch and the box isn't required.

(2) Equipment Grounding Conductor. An equipment grounding conductor or equipment bonding jumper is connected to the grounding terminal of the metal mounting yoke. **Figure 404–2**

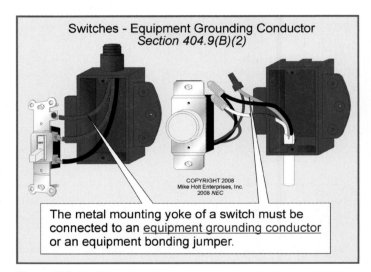

Switches - Equipment Grounding Conductor
Section 404.9(B)(2)

The metal mounting yoke of a switch must be connected to an equipment grounding conductor or an equipment bonding jumper.

Figure 404–2

Exception: The metal mounting yoke of a replacement switch isn't required to be connected to an equipment grounding conductor of a type recognized in 250.118 if the wiring to the existing switch doesn't contain an equipment grounding conductor and the switch faceplate is nonmetallic or the replacement switch is GFCI-protected.

406.5 Receptacle Faceplates. Faceplates for receptacles must completely cover the outlet openings, and they must seat firmly against the mounting surface.

(B) Grounding. Metal faceplates for receptacles must be connected to the circuit equipment grounding conductor.

> **Author's Comment:** The *NEC* doesn't specify how this is accomplished, but 517.13(B) Ex 1 for health care facilities permits the metal mounting screw(s) securing the faceplate to a metal outlet box or wiring device to be suitable for this purpose. **Figure 406–1**

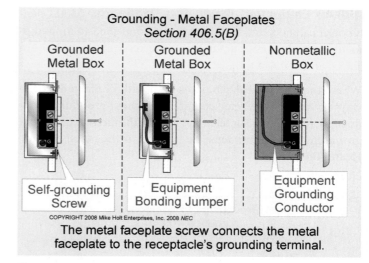

Grounding - Metal Faceplates
Section 406.5(B)

Grounded Metal Box — Self-grounding Screw

Grounded Metal Box — Equipment Bonding Jumper

Nonmetallic Box — Equipment Grounding Conductor

COPYRIGHT 2008 Mike Holt Enterprises, Inc. 2008 *NEC*

The metal faceplate screw connects the metal faceplate to the receptacle's grounding terminal.

Figure 406–1

408 Switchboards and Panelboards

408.40 Equipment Grounding Conductor. Metal panelboard cabinets and frames must be connected to an equipment grounding conductor of a type recognized in 250.118 [215.6 and 250.4(A)(3)].

Where the panelboard cabinet is used with nonmetallic raceways or cables, or where separate equipment grounding conductors are provided, a terminal bar for the circuit equipment grounding conductors must be bonded to the metal cabinet. **Figure 408–1**

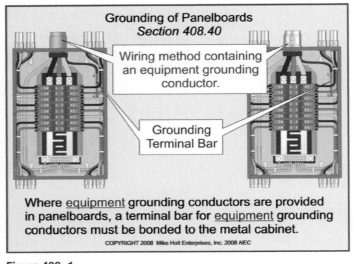

Grounding of Panelboards
Section 408.40

Wiring method containing an equipment grounding conductor.

Grounding Terminal Bar

Where equipment grounding conductors are provided in panelboards, a terminal bar for equipment grounding conductors must be bonded to the metal cabinet.

COPYRIGHT 2008 Mike Holt Enterprises, Inc. 2008 NEC

Figure 408–1

Exception: Insulated equipment grounding conductors for receptacles having insulated grounding terminals (isolated ground receptacles) [250.146(D)] can pass through the panelboard without terminating onto the equipment grounding terminal of the panelboard cabinet. **Figure 408–2**

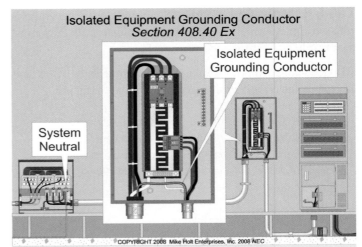

Isolated Equipment Grounding Conductor
Section 408.40 Ex

Isolated Equipment Grounding Conductor

System Neutral

COPYRIGHT 2008 Mike Holt Enterprises, Inc. 2008 NEC

An isolated equipment grounding conductor can pass through a metal enclosure, but it must terminate to the system neutral.

Figure 408–2

Equipment grounding conductors must not terminate on the neutral terminal bar, and neutral conductors must not terminate on the equipment grounding terminal bar, except as permitted by 250.142 for services and separately derived systems. **Figure 408–3**

Author's Comment: See the definition of "Separately Derived System" in Article 100.

CAUTION: *Most panelboards are rated as suitable for use as service equipment, which means they are supplied with a main bonding jumper [250.28]. This screw or strap must not be installed except when the panelboard is used for service equipment [250.24(A)(5)] or separately derived systems [250.30(A)(1)]. In addition, a panelboard marked "suitable only for use as service equipment" means the neutral bar or terminal of the panelboard has been bonded to the case at the factory, and this panelboard is restricted to being used only for service equipment or on separately derived systems according to 250.142(B).*

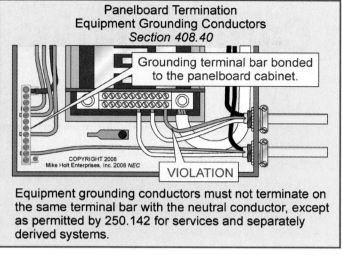

**Panelboard Termination
Equipment Grounding Conductors**
Section 408.40

Grounding terminal bar bonded to the panelboard cabinet.

COPYRIGHT 2008
Mike Holt Enterprises, Inc. 2008 *NEC*

VIOLATION

Equipment grounding conductors must not terminate on the same terminal bar with the neutral conductor, except as permitted by 250.142 for services and separately derived systems.

Figure 408–3

517.13 Grounding of Equipment in Patient Care Areas.
Wiring in patient care areas must comply with (A) and (B):

Author's Comment: Patient care areas include patient rooms as well as examining rooms, therapy areas, examining and treatment rooms, and some patient corridors. They don't include business offices, corridors, lounges, day rooms, dining rooms, or similar areas not classified as patient care areas [517.2].

(A) Wiring Methods. All branch circuits serving patient care areas must be provided with an effective ground-fault current path by installing circuits that serve patient care areas in a metal raceway or cable having a metallic armor or sheath that qualifies as an equipment grounding conductor in accordance with 250.118. **Figure 517–1**

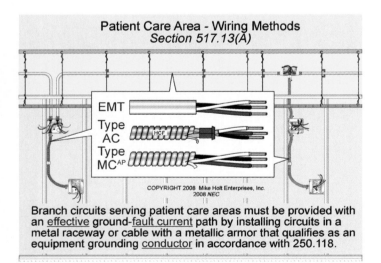

Patient Care Area - Wiring Methods
Section 517.13(A)

EMT
Type AC
Type MC^AP

COPYRIGHT 2008 Mike Holt Enterprises, Inc. 2008 NEC

Branch circuits serving patient care areas must be provided with an <u>effective</u> ground-<u>fault current</u> path by installing circuits in a metal raceway or cable with a metallic armor that qualifies as an equipment grounding <u>conductor</u> in accordance with 250.118.

Figure 517–1

Author's Comment: The metal outer sheath of AC cable is listed as an equipment grounding conductor because it contains an internal bonding strip in direct contact with the metal sheath of the cable [250.118(8)]. The metal outer sheath of interlocked Type MC cable is only listed as an equipment grounding conductor when it contains a bare aluminum conductor that makes direct contact with the metal sheath of the cable [250.118(10)(a)]. **Figure 517–2**

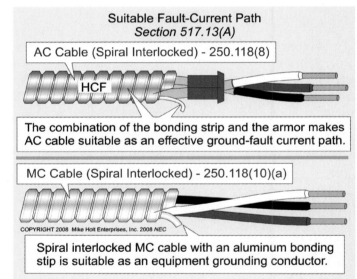

Suitable Fault-Current Path
Section 517.13(A)

AC Cable (Spiral Interlocked) - 250.118(8)

HCF

The combination of the bonding strip and the armor makes AC cable suitable as an effective ground-fault current path.

MC Cable (Spiral Interlocked) - 250.118(10)(a)

COPYRIGHT 2008 Mike Holt Enterprises, Inc. 2008 NEC

Spiral interlocked MC cable with an aluminum bonding stip is suitable as an equipment grounding conductor.

Figure 517–2

(B) Insulated Equipment Grounding Conductor. In patient care areas, the grounding terminals of receptacles and conductive surfaces of fixed electrical equipment must be connected to an insulated copper equipment grounding conductor. The circuit equipment grounding conductor must be sized in accordance with 250.122 and the wiring method must meet the requirements of 517.13(A). **Figure 517–3**

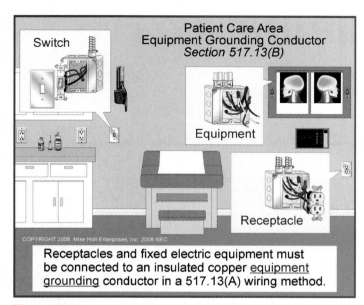

Figure 517–3

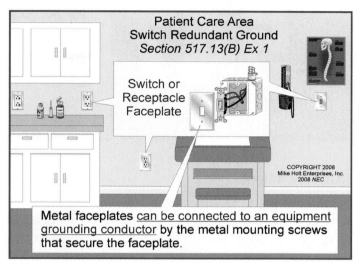

Figure 517–4

Exception No. 1: Metal faceplates for switches and receptacles can be connected to the equipment grounding conductor by the metal mounting screws that secure the faceplate to a metal outlet box or metal mounting yoke of switches [404.9(B)] and receptacles [406.3(C)]. **Figure 517–4**

Exception No. 2: Luminaires located more than 7½ ft above the floor can be connected to the equipment grounding return path complying with 517.13(A), without being connected to an insulated equipment grounding conductor.

547 Agricultural Buildings

547.2 Definitions.

Equipotential Plane. An equipotential plane is an area where wire mesh or other conductive elements are embedded in or placed under concrete, and are bonded to the electrical system to prevent a voltage difference from developing within the plane [547.10(B)]. **Figure 547–1**

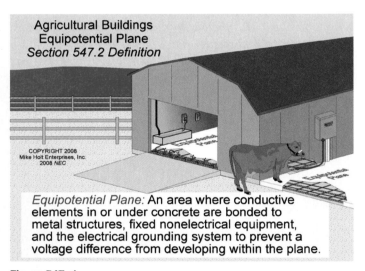

Equipotential Plane: An area where conductive elements in or under concrete are bonded to metal structures, fixed nonelectrical equipment, and the electrical grounding system to prevent a voltage difference from developing within the plane.

Figure 547–1

547.10 Equipotential Planes and Bonding of Equipotential Planes. The installation and bonding of equipotential planes must comply with (A) and (B).

(A) Where Equipotential Plane is Required. An equipotential plane must be installed in:

(1) Indoors. Indoor concrete floor confinement areas where metallic equipment is located that may become energized and is accessible to livestock. **Figure 547–2**

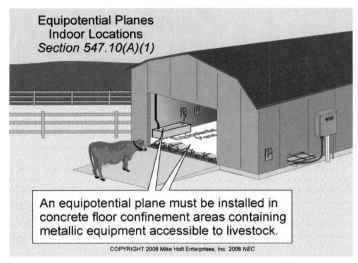

An equipotential plane must be installed in concrete floor confinement areas containing metallic equipment accessible to livestock.

Figure 547–2

(2) Outdoors. Outdoor concrete confinement areas where metallic equipment is located that may become energized and is accessible to livestock. **Figure 547–3**

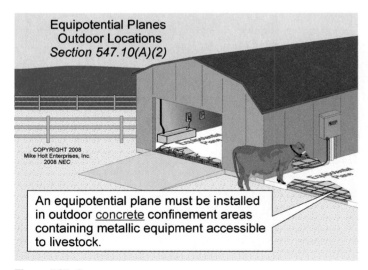

An equipotential plane must be installed in outdoor concrete confinement areas containing metallic equipment accessible to livestock.

Figure 547–3

The equipotential plane must encompass the area around the equipment where the livestock stand.

(B) Bonding of Equipotential Plane. The equipotential plane must be connected to the building or structure's electrical grounding system. The bonding conductor used for this purpose must be insulated, covered, or bare copper and not smaller than 8 AWG. The bonding conductor must terminate at pressure connectors or clamps of brass, copper, copper alloy, or an equally substantial means approved by the authority having jurisdiction [250.8].

Author's Comment: See the definition of "Connector, Pressure" in Article 100.

FPN No. 1: Methods to establish equipotential planes are described in American Society of Agricultural and Biological Engineers Standard EP473, *Equipotential Planes in Animal Containment Areas.*

Author's Comment: This standard provides the recommendation of a voltage gradient ramp at the entrances of agricultural buildings. **Figure 547–4**

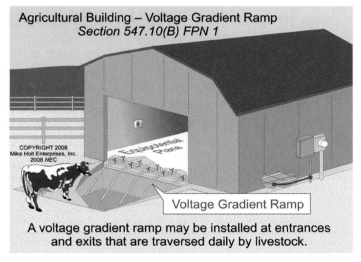

A voltage gradient ramp may be installed at entrances and exits that are traversed daily by livestock.

Figure 547–4

FPN No. 2: See American Society of Agricultural and Biological Engineers EP342.2, *Safety for Electrically Heated Livestock Waterers.*

Author's Comment: The bonding requirements contained in Article 547 are unique because of the sensitivity of livestock to stray voltage/current, especially in wet or damp concrete animal confinement areas.

In most instances the voltage difference between metal parts and the earth will be too low to present a shock hazard to persons. However, livestock might detect the voltage difference if they come in contact with the metal parts. Although potential differences may not be life threatening to the livestock, it has been reported that as little as 0.50V RMS can affect milk production.

The topic of stray voltage/current is beyond the scope of this textbook. For more information, visit www.MikeHolt.com. Click on the "Technical" link, then the "Stray Voltage" link for more details.

600.7 Grounding and Bonding.

(A) Grounding.

(1) Equipment Grounding. Signs and metal equipment of outline lighting systems must be connected to the circuit equipment grounding conductor of the supply circuit using an equipment grounding conductor recognized in 250.118. Figure 600–1

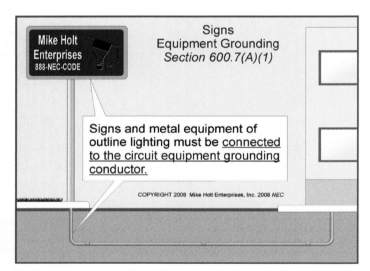

Signs
Equipment Grounding
Section 600.7(A)(1)

Mike Holt
Enterprises
888-NEC-CODE

Signs and metal equipment of outline lighting must be <u>connected to the circuit equipment grounding conductor.</u>

COPYRIGHT 2008 Mike Holt Enterprises, Inc. 2008 NEC

Figure 600–1

(2) Size of Equipment Grounding Conductor. The circuit equipment grounding conductor size must be in accordance with 250.122, based on the rating of the overcurrent device protecting the conductors supplying the sign.

(3) Connections. Equipment grounding conductor connections must be made in accordance with 250.8.

> **Author's Comment:** According to 250.8, equipment grounding conductors must terminate in one of the following methods:
>
> (1) Listed pressure connectors
> (2) Terminal bars

(3) Pressure connectors listed for direct burial or concrete encasement [250.70]
(4) Exothermic welding
(5) Machine screws that engage at least two threads or are secured with a nut
(6) Self-tapping machine screws that engage at least two threads
(7) Connections that are part of a listed assembly
(8) Other listed means

(4) Auxiliary Grounding Electrode. Auxiliary grounding electrodes are not required for signs, but if installed, they must comply with 250.54.

> **Author's Comment:** According to 250.54, auxiliary electrodes are permitted and they need not be bonded to the building or structure grounding electrode system, the grounding conductor to the electrode need not be sized according to 250.66, and the contact resistance of the electrode to the earth is not required to comply with the 25 ohm requirement of 250.56. **Figure 600–2**
>
> The earth must not be used as the effective ground-fault current path required by 250.4(A)(4). This is because the contact resistance of a grounding electrode to the earth is high, and very little ground-fault current returns to the electrical supply source via the earth. The result is the circuit overcurrent device will not open and clear a ground fault; therefore, metal parts will remain energized with dangerous voltage.

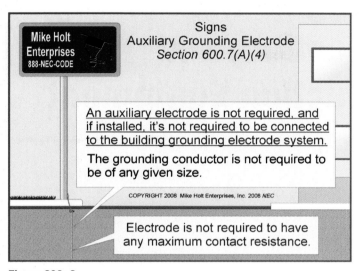

Figure 600–2

ARTICLE 640

Audio Signal Processing, Amplification, and Reproduction Equipment

PART II. PERMANENT AUDIO SYSTEM INSTALLATIONS

640.22 Wiring of Equipment Racks. Metal equipment racks and enclosures must be connected to an equipment grounding conductor of a type recognized in 250.118 [250.4(A)(3)].

645 Information Technology Equipment

645.15 Equipment Grounding Conductor. Signal reference structures must be bonded to the circuit equipment grounding conductor for the information technology equipment. **Figures 645–1 and 645–2**

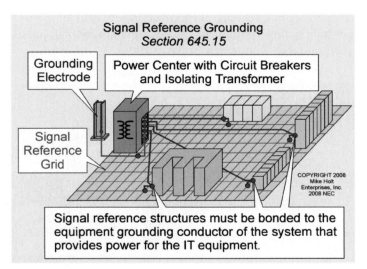

Figure 645–1

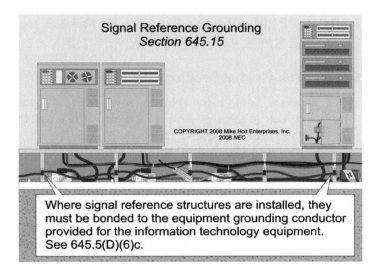

Figure 645–2

FPN No. 2: Where isolated ground receptacles are installed, they must be connected to an insulated equipment grounding conductor in accordance with 250.146(D) and 406.2(D).

PART II. PERMANENTLY INSTALLED POOLS, OUTDOOR SPAS, AND OUTDOOR HOT TUBS

680.26 Equipotential Bonding.

(A) Performance. Equipotential bonding is intended to reduce voltage gradients in the area around a permanently installed pool, outdoor spa, or outdoor hot tub by the use of a common bonding grid in accordance with 680.26(B) and (C).

(B) Bonded Parts. The parts of a permanently installed pool, outdoor spa, or outdoor hot tub listed in (B)(1) through (B)(7) must be bonded together with a solid copper conductor not smaller than 8 AWG with listed pressure connectors, terminal bars, exothermic welding, or other listed means [250.8(A)]. **Figure 680-1**

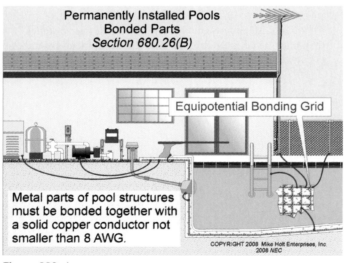

Figure 680-1

Equipotential bonding is not required to extend to or be attached to any panelboard, service equipment, or grounding electrode.

(1) Concrete Pool, Outdoor Spa, and Outdoor Hot Tub Shells.

(a) Structural Reinforcing Steel. Unencapsulated structural reinforcing steel in concrete shells secured together by steel tie wires.

(2) Perimeter Surfaces. An equipotential bonding grid must extend 3 ft horizontally beyond the inside walls of a pool, outdoor spa, or outdoor hot tub, including unpaved, paved, and poured concrete surfaces. **Figure 680-2**

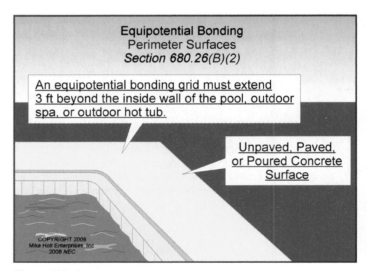

Figure 680-2

The bonding grid must comply with (a) or (b) and be attached to the conductive pool reinforcing steel at a minimum of four points uniformly spaced around the perimeter of the walls of a pool, outdoor spa, or outdoor hot tub.

(a) Structural Reinforcing Steel. Structural reinforcing steel [680.26(B)(1)(a)]. **Figure 680-3**

> **Author's Comment:** The 2008 *NEC* does not provide any guidance on the installation requirements for structural reinforcing steel when used as a perimeter equipotential bonding grid.

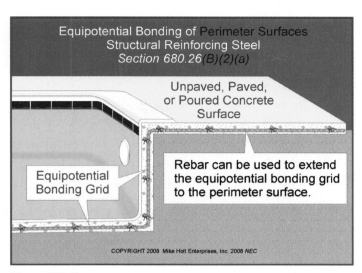

Equipotential Bonding of Perimeter Surfaces
Structural Reinforcing Steel
Section 680.26(B)(2)(a)

Unpaved, Paved, or Poured Concrete Surface

Equipotential Bonding Grid

Rebar can be used to extend the equipotential bonding grid to the perimeter surface.

COPYRIGHT 2008 Mike Holt Enterprises, Inc. 2008 NEC

Figure 680–3

(b) Alternate Means. Equipotential bonding conductor meeting all of the following requirements: **Figure 680–4**

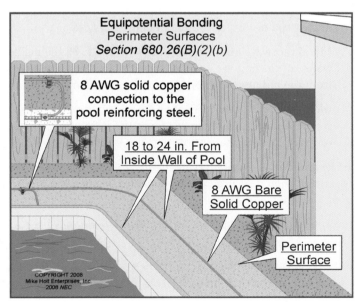

Equipotential Bonding
Perimeter Surfaces
Section 680.26(B)(2)(b)

8 AWG solid copper connection to the pool reinforcing steel.

18 to 24 in. From Inside Wall of Pool

8 AWG Bare Solid Copper

Perimeter Surface

COPYRIGHT 2008 Mike Holt Enterprises, Inc. 2008 NEC

Figure 680–4

(1) 8 AWG bare solid copper bonding conductor.

(2) The bonding conductor must follow the contour of the perimeter surface.

(3) Listed splicing devices.

(4) Bonding conductor must be 18 to 24 in. from the inside walls of the pool.

(5) Bonding conductor must be secured within or under the perimeter surface 4 to 6 in. below the subgrade.

(3) Metallic Components. Metallic parts of the pool, outdoor spa, or outdoor hot tub structure must be bonded to the equipotential grid.

(4) Underwater Metal Forming Shells. Metal forming shells and mounting brackets for luminaires and speakers must be bonded to the equipotential grid.

(5) Metal Fittings. Metal fittings sized 4 in. and larger that penetrate into the pool, outdoor spa, or outdoor hot tub structure, such as ladders and handrails must be bonded to the equipotential grid.

(6) Electrical Equipment. Metal parts of electrical equipment associated with the pool, outdoor spa, or outdoor hot tub water circulating system, such as water heaters and pump motors and metal parts of pool covers must be bonded to the equipotential grid. **Figure 680–5**

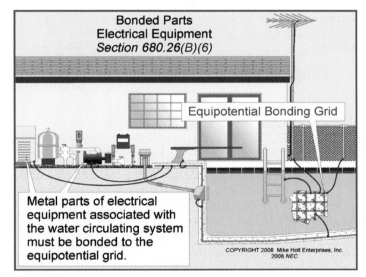

Bonded Parts
Electrical Equipment
Section 680.26(B)(6)

Equipotential Bonding Grid

Metal parts of electrical equipment associated with the water circulating system must be bonded to the equipotential grid.

COPYRIGHT 2008 Mike Holt Enterprises, Inc. 2008 NEC

Figure 680–5

Exception: Metal parts of listed equipment incorporating an approved system of double insulation are not required to be bonded to the equipotential grid.

(a) Double-Insulated Water Pump Motors. Where a double-insulated water pump motor is installed, a solid 8 AWG copper conductor from the bonding grid must be provided for a replacement motor.

(b) Pool Water Heaters. Pool water heaters must be grounded and bonded in accordance with equipment instructions.

(7) Metal Wiring Methods and Equipment. Metal-sheathed cables and raceways, metal piping, and all fixed metal parts must be bonded to the equipotential grid.

Exception No. 1: Where separated from the pool, outdoor spa, or outdoor hot tub structure by a permanent barrier.

Exception No. 2: Where located more than 5 ft horizontally of the inside walls of the pool, outdoor spa, or outdoor hot tub structure.

Exception No. 3: Where located more than 12 ft measured vertically above the maximum water level.

(C) Pool Water. A minimum conductive surface area of 9 sq in. must be installed in contact with the pool, outdoor spa, or outdoor hot tub structure water. This water bond is permitted to consist of metal parts that are bonded in 680.26(B). Figure 680–6

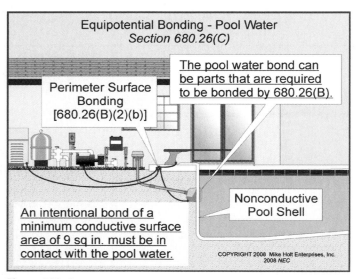

Figure 680–6

PART IV. SPAS AND HOT TUBS

680.42 Outdoor Installations. Electrical installations for outdoor spas or hot tubs must comply with Parts I and II of this article, except as permitted for (B) or (C).

(B) Bonding. Bonding is permitted by mounting equipment to a metal frame or base. Metal bands that secure wooden staves aren't required to be bonded.

(C) Interior Wiring for Outdoor Spas or Hot Tubs. Any Chapter 3 wiring method containing a copper equipment grounding conductor insulated or enclosed within the outer sheath of the wiring method and not smaller than 12 AWG is permitted in the interior of a one-family dwelling for the connection to motor, heating, and control loads that are part of a self-contained spa or hot tub, or a packaged spa or hot tub equipment assembly.

Wiring to an underwater light must comply with 680.23 or 680.33.

680.43 Indoor Installations. Electrical installations for an indoor spa or hot tub must comply with Parts I and II of Article 680, except as modified by this section. Indoor installations of spas or hot tubs can be connected by any of the wiring methods contained in Chapter 3.

(D) Bonding. The following parts of an indoor spa or hot tub must be bonded together:

(1) Metal fittings within or attached to the indoor spa or hot tub structure.

(2) Metal parts of electrical equipment associated with the indoor spa or hot tub water circulating system.

(3) Metal raceways and metal piping within 5 ft of the inside walls of the indoor spa or hot tub, and not separated from the indoor spa or hot tub by a permanent barrier.

(4) Metal surfaces within 5 ft of the inside walls of an indoor spa or hot tub not separated from the indoor spa or hot tub area by a permanent barrier.

Exception No. 1: Nonelectrical equipment, such as towel bars or mirror frames, which aren't connected to metallic piping, aren't required to be bonded.

Exception No. 2: Metal parts of a listed self-contained spa or hot tub.

(E) Methods of Bonding. Metal parts associated with the spa or hot tub as described in 680.43(D) must be bonded by any of the following methods:

(1) Threaded metal piping and fittings.

(2) Metal-to-metal mounting to a common frame or base.

(3) A solid copper conductor not smaller than 8 AWG.

PART VII. HYDROMASSAGE BATHTUBS

680.74 Equipotential Bonding. Metal piping in contact with the circulating water of the hydromassage bathtub must be bonded, with a solid copper conductor not smaller than 8 AWG, to the circulating pump if it's not double insulated.

The equipotential hydromassage bonding jumper is not required to be bonded to any remote panelboard, service equipment, or electrode. **Figure 680–7**

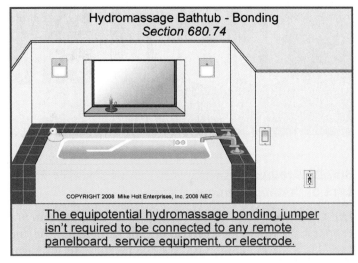

Hydromassage Bathtub - Bonding
Section 680.74

COPYRIGHT 2008 Mike Holt Enterprises, Inc. 2008 *NEC*

The equipotential hydromassage bonding jumper isn't required to be connected to any remote panelboard, service equipment, or electrode.

Figure 680–7

ARTICLE 800 Communications Circuits

800.93 Grounding or Interruption of Metallic Members of Communications Cables.

(A) Communications Cables Entering Building. Communications cables entering the building or terminating on the outside of the building must have the metallic sheath members either grounded as specified in 800.100, or interrupted by an insulating joint as close as practicable to the point of entrance. **Figure 800–1**

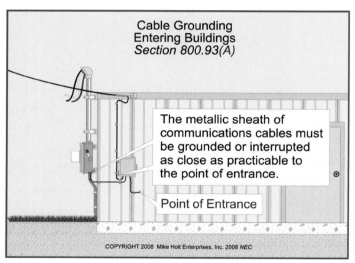

Cable Grounding
Entering Buildings
Section 800.93(A)

The metallic sheath of communications cables must be grounded or interrupted as close as practicable to the point of entrance.

Point of Entrance

COPYRIGHT 2008 Mike Holt Enterprises, Inc. 2008 *NEC*

Figure 800–1

FPN: The point of entrance is defined as the point within a building at which the communications cable emerges from an external wall, from a concrete floor slab, or from a rigid metal conduit or an intermediate metal conduit connected by a grounding conductor to an electrode in accordance with 800.100 [800.2].

Author's Comment: Limiting the length of the grounding conductor helps limit damage to equipment because of a potential (voltage) difference between communications equipment and other systems during lightning events [250.(4)(A)(1) FPN].

PART IV. GROUNDING METHODS

800.100 Cable Grounding. The primary protector and the metallic sheath of communications cables must be grounded in accordance with the following requirements:

(A) Grounding Conductor. The grounding conductor must be:

(1) Insulation. The grounding conductor must be insulated and must be listed.

(2) Material. The grounding conductor must be copper or other corrosion-resistant conductive material, stranded or solid.

(3) Size. The grounding conductor must not be smaller than 14 AWG.

(4) Length. The grounding conductor must be as short as practicable. In one- and two-family dwellings, the grounding conductor must not exceed 20 ft. **Figure 800–2**

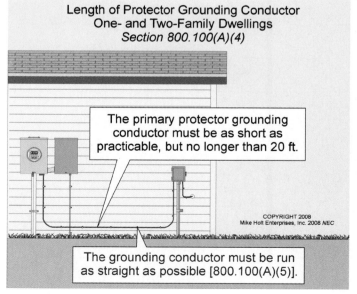

Length of Protector Grounding Conductor
One- and Two-Family Dwellings
Section 800.100(A)(4)

The primary protector grounding conductor must be as short as practicable, but no longer than 20 ft.

COPYRIGHT 2008
Mike Holt Enterprises, Inc. 2008 *NEC*

The grounding conductor must be run as straight as possible [800.100(A)(5)].

Figure 800–2

FPN: Limiting the length of the grounding conductor helps limit induced potential (voltage) differences between the building's power and communications systems during lightning events.

Exception: Where the grounding conductor is over 20 ft for one- and two-family dwellings, a separate ground rod not less than 5 ft long [800.100(B)(3)(2)] with fittings suitable for the application [800.100(C)] must be installed. The additional ground rod must be bonded to the power grounding electrode system with a minimum 6 AWG conductor [800.100(D)].

(5) Run in Straight Line. The grounding conductor to the electrode must be run in as straight a line as practicable.

Author's Comment: Lightning doesn't like to travel around corners or through loops, which is why the grounding conductor must be run as straight as practicable.

(6) Physical Protection. The grounding conductor must be mechanically protected where subject to physical damage, and where run in a metal raceway both ends of the raceway must be bonded to the grounding conductor.

Author's Comment: Installing the grounding conductor in PVC conduit is a better practice.

(B) Electrode. The grounding conductor must be connected in accordance with (B)(1), (B)(2), or (B)(3):

(1) Buildings or Structures with an Intersystem Bonding Termination. The grounding conductor for the primary protector and the metallic sheath of communications cable must terminate to the intersystem bonding terminal [Article 100 and 250.94]. **Figure 800–3**

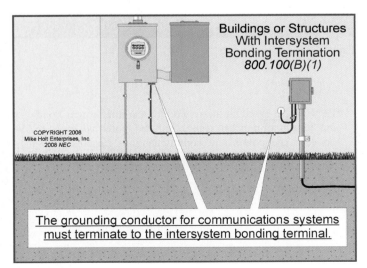

Buildings or Structures With Intersystem Bonding Termination *800.100(B)(1)*

The grounding conductor for communications systems must terminate to the intersystem bonding terminal.

Figure 800–3

Author's Comment: The bonding of all external communications systems to a single point minimizes the possibility of damage because of potential (voltage) differences between the systems. **Figure 800–4**

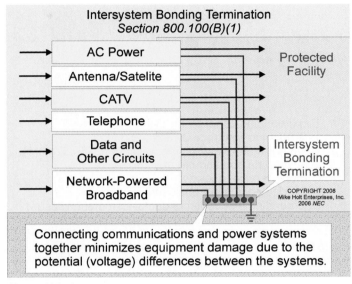

Intersystem Bonding Termination *Section 800.100(B)(1)*

AC Power → Protected Facility

Antenna/Satelite

CATV

Telephone

Data and Other Circuits

Network-Powered Broadband

Intersystem Bonding Termination

Connecting communications and power systems together minimizes equipment damage due to the potential (voltage) differences between the systems.

Figure 800–4

(2) Building or Structure Without Intersystem Bonding Termination. The grounding conductor must terminate to the nearest accessible: **Figure 800–5**

(1) Building or structure grounding electrode system [250.50].

(2) Interior metal water piping system, within 5 ft from its point of entrance [250.52(A)(1)].

(3) Accessible means external to the building, as covered in 250.94.

(4) Metallic service raceway.

(5) Service equipment enclosure.

(6) Grounding electrode conductor or the grounding electrode conductor metal enclosure.

(7) The grounding conductor or the grounding electrode of a remote building or structure disconnecting means [250.32].

The intersystem bonding terminal must be mounted on the fixed part of an enclosure so that it will not interfere with the opening of an enclosure door. A bonding device must not be mounted on a door or cover even if the door or cover is non-removable.

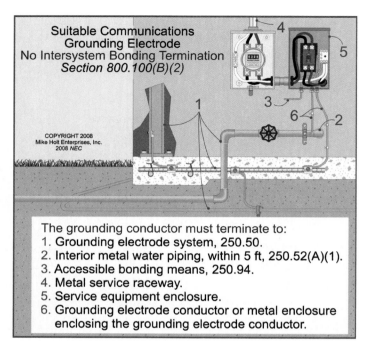

Figure 800–5

(3) In Buildings or Structures Without Intersystem Bonding Termination or Grounding Means. The grounding conductor must connect to:

(1) To any individual electrodes described in 250.52(A)(1), (A)(2), (A)(3) or (A)(4).

(2) To any individual electrodes described in 250.52(A)(6) and (A)(7), or to a ground rod not less than 5 ft long and ½ in. in diameter. **Figure 800–6**

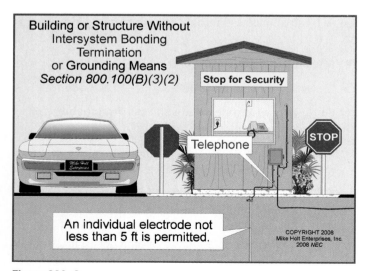

Figure 800–6

Author's Comment: The reason communications ground rods only need to be 5 ft long is because that's the length the phone company used before the *NEC* contained requirements for communications systems. Phone company ground rods were only 5 ft long because that's the length that would fit in their equipment trailers.

(C) Electrode Connection. Terminations at the electrode must be by exothermic welding, listed lugs, listed pressure connectors, or listed clamps. Grounding fittings that are concrete-encased or buried in the earth must be listed for direct burial [250.70].

(D) Bonding of Electrodes. If a separate grounding electrode, such as a ground rod, is installed for a communications system, it must be bonded to the building's power grounding electrode system with a minimum 6 AWG conductor. **Figure 800–7**

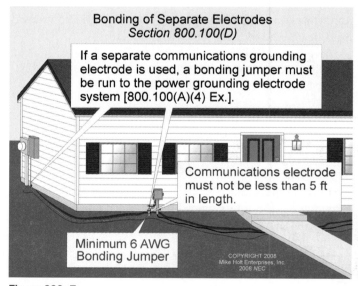

Figure 800–7

FPN No. 2: Bonding all systems to the intersystem bonding termination helps reduce induced potential (voltage) between the power and communications systems during lightning events. **Figure 800–8**

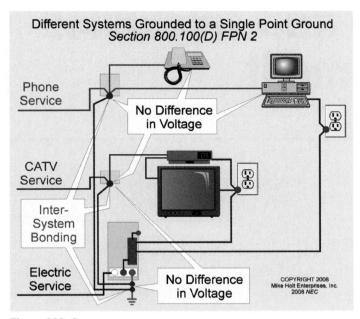

Different Systems Grounded to a Single Point Ground
Section 800.100(D) FPN 2

Phone Service

No Difference in Voltage

CATV Service

Inter-System Bonding

Electric Service

No Difference in Voltage

COPYRIGHT 2008
Mike Holt Enterprises, Inc.
2008 *NEC*

Figure 800–8

PART II. RECEIVING EQUIPMENT— ANTENNA SYSTEMS

810.15 Metal Antenna Supports—Grounding. Outdoor masts and metal structures that support antennas must be grounded in accordance with 810.21.

810.20 Antenna Discharge Unit.

(A) Required. Each lead-in conductor from an outdoor antenna must be provided with a listed antenna discharge unit.

(B) Location. The antenna discharge unit must be located outside or inside the building, nearest the point of entrance, but not near combustible material.

(C) Grounding. The antenna discharge unit must be grounded in accordance with 810.21.

810.21 Grounding Conductors. The antenna mast [810.15] and antenna discharge unit [810.20(C)] must be grounded as follows. **Figure 810–1**

> **Author's Comment:** Grounding the lead-in antenna cables and the mast help prevent voltage surges caused by static discharge or nearby lightning strikes from reaching the center conductor of the lead-in coaxial cable. Because the satellite sits outdoors, wind creates a static charge on the antenna as well as on the cable attached to it. This charge can build up on both the antenna and the cable until it jumps across an air space, often passing through the electronics inside the low noise block down converter feed horn (LNBF) or receiver. Connecting the coaxial cable and dish to the building grounding electrode system (grounding) helps to dissipate this static charge.
>
> Nothing can prevent damage from a direct lightning strike. But grounding with proper surge protection can help reduce damage to satellite and other equipment from nearby lightning strikes.

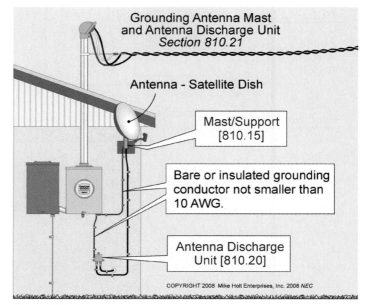

Figure 810–1

(A) Material. The grounding conductor to the electrode [810.21(F)] must be copper or other corrosion-resistant conductive material, stranded or solid.

(B) Insulation. The grounding conductor isn't required to be insulated.

(C) Supports. The grounding conductor must be securely fastened in place.

(D) Mechanical Protection. The grounding conductor must be mechanically protected where subject to physical damage, and where run in a metal raceway both ends of the raceway must be bonded to the grounding conductor. **Figure 810–2**

> **Author's Comment:** Installing the grounding conductor in PVC conduit is a better practice.

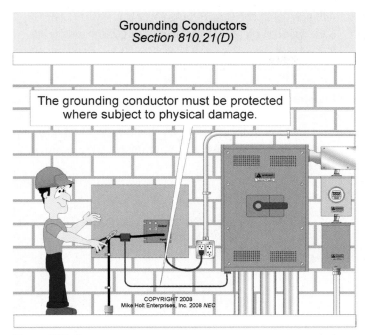

Grounding Conductors
Section 810.21(D)

The grounding conductor must be protected where subject to physical damage.

COPYRIGHT 2008
Mike Holt Enterprises, Inc. 2008 NEC

Figure 810–2

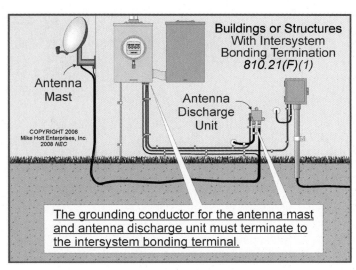

Buildings or Structures
With Intersystem
Bonding Termination
810.21(F)(1)

Antenna Mast

Antenna Discharge Unit

COPYRIGHT 2008
Mike Holt Enterprises, Inc.
2008 NEC

The grounding conductor for the antenna mast and antenna discharge unit must terminate to the intersystem bonding terminal.

Figure 810–3

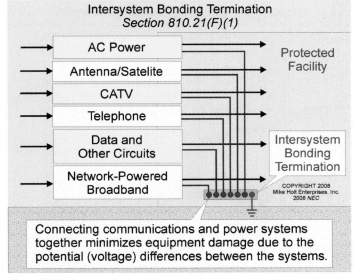

Intersystem Bonding Termination
Section 810.21(F)(1)

AC Power

Antenna/Satelite

CATV

Telephone

Data and Other Circuits

Network-Powered Broadband

Protected Facility

Intersystem Bonding Termination

COPYRIGHT 2008
Mike Holt Enterprises, Inc.
2008 NEC

Connecting communications and power systems together minimizes equipment damage due to the potential (voltage) differences between the systems.

Figure 810–4

(E) Run in Straight Line. The grounding conductor must be run in as straight a line as practicable.

Author's Comment: Lightning doesn't like to travel around corners or through loops, which is why the grounding conductor must be run as straight as practicable.

(F) Electrode.

(1) Buildings or Structures with an Intersystem Bonding Termination. The grounding conductor for the antenna mast and antenna discharge unit must terminate to the intersystem bonding terminal [Article 100 and 250.94]. **Figure 810–3**

Author's Comment: Bonding all systems to the intersystem bonding termination helps reduce induced potential (voltage) differences between the power and radio and television systems during lightning events. **Figure 810–4**

(2) In Buildings or Structures without Intersystem Bonding Termination. The grounding conductor for the antenna mast and antenna discharge unit must terminate to the nearest accessible: **Figure 800–5**

(1) Building or structure grounding electrode system [250.50]

(2) Interior metal water piping system, within 5 ft from its point of entrance [250.52(A)(1)], **Figure 800–6**

(3) Accessible means external to the building, as covered in 250.94

(4) Metallic service raceway

(5) Service equipment enclosure

(6) Grounding electrode conductor or the grounding electrode conductor metal enclosure

(3) In Buildings or Structures Without a Grounding Means. The grounding conductor must connect to:

(1) To any individual electrodes described in 250.52

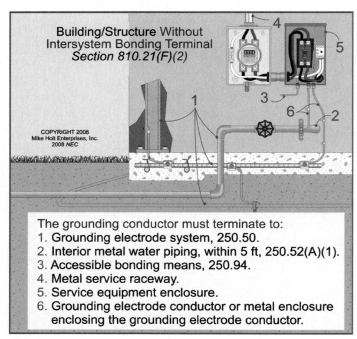

Figure 810–5

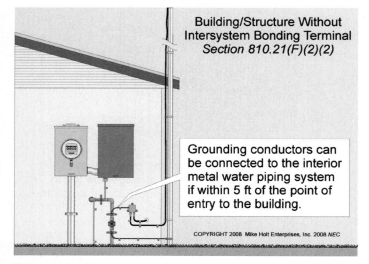

Figure 810–6

(2) To structural steel grounded in accordance with 250.52(A)(2)

(G) Inside or Outside Building. The grounding conductor can be run either inside or outside the building.

(H) Size. The grounding conductor must not be smaller than 10 AWG copper or 17 AWG copper-clad steel or bronze.

Author's Comment: Copper-clad steel or bronze wire (17 AWG) is often molded into the jacket of the coaxial cable to simplify the grounding of the satellite dish by eliminating the need to run a separate grounding conductor to the dish [810.21(F)(2)].

(J) Bonding of Electrodes. If a ground rod is installed to serve as the grounding electrode for the radio and television equipment, it must be connected to the building's power grounding electrode system with a minimum 6 AWG conductor. **Figure 810–7**

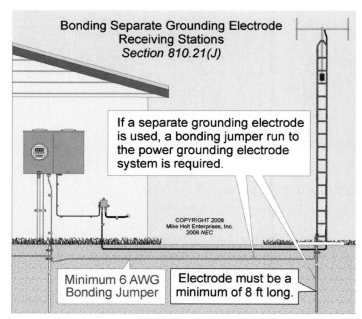

Figure 810–7

(K) Electrode Connection. Termination of the grounding conductor must be by exothermic welding, listed lugs, listed pressure connectors, or listed clamps. Grounding fittings that are concrete-encased or buried in the earth must be listed for direct burial [250.70]. **Figure 810–8**

PART III. AMATEUR TRANSMITTING AND RECEIVING STATIONS—ANTENNA SYSTEMS

810.51 Other Sections. Antenna systems for amateur transmitting and receiving stations must also comply with the following requirements:

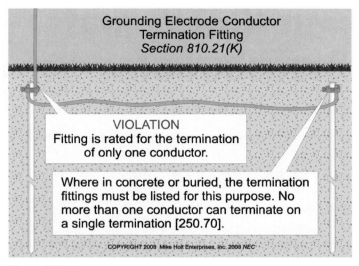

Grounding Electrode Conductor
Termination Fitting
Section 810.21(K)

VIOLATION
Fitting is rated for the termination
of only one conductor.

Where in concrete or buried, the termination
fittings must be listed for this purpose. No
more than one conductor can terminate on
a single termination [250.70].

COPYRIGHT 2008 Mike Holt Enterprises, Inc. 2008 NEC

Figure 810–8

Support of Lead-In Cables. Antennas and lead-in conductors must be securely supported, and the lead-in conductors must be securely attached to the antenna [810.12].

Avoid Contact with Conductors of Other Systems. Outdoor antennas and lead-in conductors must be kept at least 2 ft from exposed electric power conductors to avoid the possibility of accidental contact [810.13].

Metal Antenna Supports—Grounding. Outdoor masts and metal structures that support antennas must be grounded in accordance with 810.21 [810.15].

810.58 Grounding Conductors.

(A) Other Sections. The antenna mast [810.15] and antenna discharge unit [810.57] must be grounded as specified in 810.21.

(B) Size of Protective Grounding Conductor. The grounding conductor must be the same size as the lead-in conductors, but not smaller than 10 AWG copper, bronze, or copper-clad steel.

(C) Size of Operating Grounding Conductor. The grounding conductor for transmitting stations must not be smaller than 14 AWG copper or its equivalent.

ARTICLE 820
Community Antenna Television (CATV) and Radio Distribution Systems

PART III. PROTECTION

820.93 Grounding or Interruption of Metallic Members of Coaxial CATV Cables.

(A) Coaxial CATV Cables Entering Building. Coaxial CATV cables entering the building or terminating on the outside of the building must have the metallic sheath members either grounded as specified in 820.100, or interrupted by an insulating joint as close as practicable to the point of entrance. **Figure 820–1**

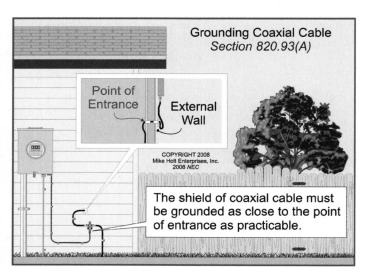

Grounding Coaxial Cable
Section 820.93(A)

Point of Entrance

External Wall

COPYRIGHT 2008
Mike Holt Enterprises, Inc.
2008 NEC

The shield of coaxial cable must be grounded as close to the point of entrance as practicable.

Figure 820–1

Author's Comment: Limiting the length of the grounding conductor helps limit damage to equipment because of a potential (voltage) difference between communications equipment and other systems during lightning events [250.(4)(A)(1) FPN].

PART IV. GROUNDING METHODS

820.100 Cable Grounding. The outer conductive shield of a coaxial cable must be grounded in accordance with the following requirements:

(A) Grounding Conductor.

(1) Insulation. The grounding conductor must be insulated and must be listed.

(2) Material. The grounding conductor must be copper or other corrosion-resistant conductive material, stranded or solid.

(3) Size. The grounding conductor must not be smaller than 14 AWG, and is not required to be larger than 6 AWG. It must have a current-carrying capacity equal to the outer conductor of the coaxial cable.

(4) Length. The grounding conductor must be as short as practicable.

In one- and two-family dwellings, the grounding conductor must not exceed 20 ft. **Figure 820–2**

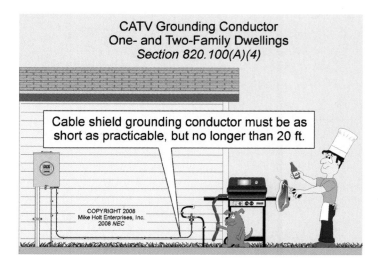

CATV Grounding Conductor
One- and Two-Family Dwellings
Section 820.100(A)(4)

Cable shield grounding conductor must be as short as practicable, but no longer than 20 ft.

COPYRIGHT 2008
Mike Holt Enterprises, Inc.
2008 NEC

Figure 820–2

FPN: Limiting the length of the grounding conductor will help to reduce potential (voltage) differences between the building's power and CATV systems during lightning events.

Exception: Where it's not practicable to limit the coaxial grounding conductor to 20 ft for one- and two-family dwellings, a separate ground rod not less than 8 ft long, with fittings suitable for the application [250.70 and 820.100(C)] must be installed. The additional ground rod must be bonded to the power grounding electrode system with a minimum 6 AWG conductor [820.100(D)]. **Figure 820–3**

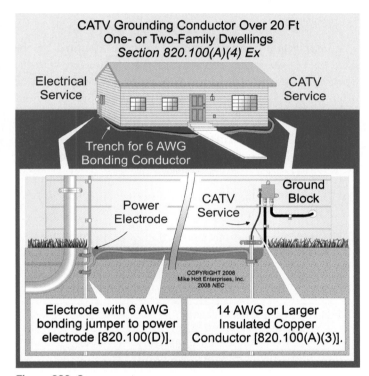

Figure 820–3

(5) Run in Straight Line. The grounding conductor to the electrode must be run in as straight a line as practicable.

Author's Comment: Lightning doesn't like to travel around corners or through loops, which is why the grounding conductor must be run as straight as practicable.

(6) Physical Protection. The grounding conductor must be mechanically protected where subject to physical damage, and where run in a metal raceway both ends of the raceway must be bonded to the grounding conductor. **Figure 820–4**

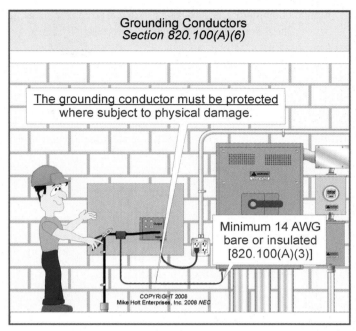

Figure 820–4

Author's Comment: Installing the grounding conductor in PVC conduit is a better practice.

(B) Electrode. The grounding conductor must be connected in accordance with (B)(1), (B)(2), or (B)(3).

(1) Buildings or Structures With an Intersystem Bonding Termination. The grounding conductor for the CATV system must terminate to the intersystem bonding terminal. **Figure 820–5**

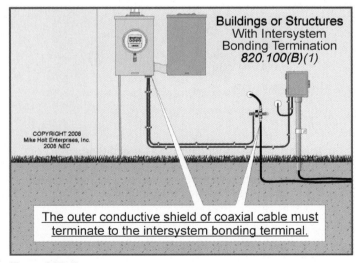

Figure 820–5

Author's Comment: The bonding of all external communications systems to a single point minimizes the possibility of damage because of potential (voltage) differences between the systems. **Figure 820–6**

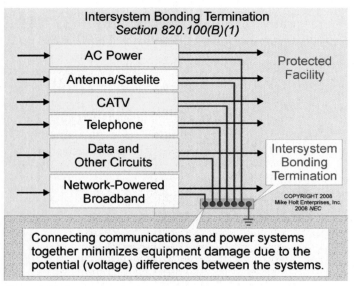

Figure 820–6

(2) In Buildings or Structures With a Grounding Means. At existing structures, the grounding conductor must terminate to the nearest accessible: **Figure 820–7**

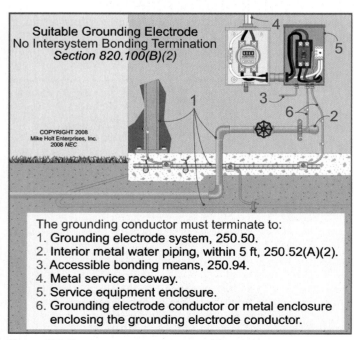

Figure 820–7

(1) Building or structure grounding electrode system [250.50]

(2) Interior metal water piping system, within 5 ft from its point of entrance [250.52(A)(1)]

(3) Accessible means external to the building, as covered in 250.94

(4) Metallic service raceway

(5) Service equipment enclosure

(6) Grounding electrode conductor or the grounding electrode conductor metal enclosure

(7) The grounding conductor or the grounding electrode of a remote building or structure disconnecting means [250.32]

The intersystem bonding terminal must be mounted on the fixed part of an enclosure so that it will not interfere with the opening of an enclosure door. A bonding device must not be mounted on a door or cover even if the door or cover is non-removable.

(3) In Buildings or Structures Without Intersystem Bonding Termination or Grounding Means. The grounding conductor must connect to:

(1) To any individual electrodes described in 250.52(A)(1), (A)(2), (A)(3) or (A)(4)

(2) To any individual electrodes described in 250.52(A)(5), 250.52(A)(6) and (A)(7). **Figure 820–8**

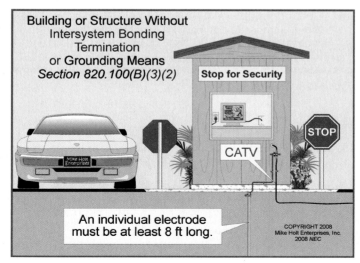

Figure 820–8

(C) Electrode Connection. Terminations to the electrode must be by exothermic welding, listed lugs, listed pressure connectors, or clamps. Grounding fittings that are concrete-encased or buried in the earth must be listed for direct burial [250.70].

(D) Bonding of Electrodes. If a separate grounding electrode, such as a ground rod, is installed for the CATV system, it must be bonded to the building's power grounding electrode system with a minimum 6 AWG conductor. **Figure 820–9**

> **FPN No. 2:** Bonding all systems to the intersystem bonding termination helps reduce induced potential (voltage) between the power and CATV system during lightning events.

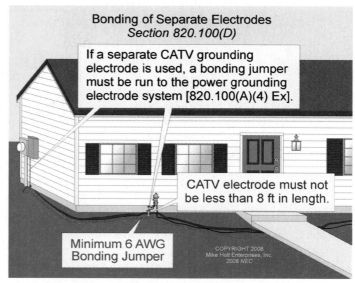

Figure 820–9

ARTICLES 300–820 Practice Questions

Use the 2008 *NEC* to answer the following questions.

NEC ARTICLES 300 THROUGH 820 PRACTICE QUESTIONS

1. Metal raceways, cable armors, and other metal enclosures shall be _____ joined together into a continuous electric conductor so as to provide effective electrical continuity.

   (a) electrically
   (b) permanently
   (c) metallically
   (d) none of these

2. Metal boxes shall be _____ in accordance with Article 250.

   (a) grounded
   (b) bonded
   (c) a and b
   (d) none of these

3. In completed installations, each outlet box shall have a _____.

   (a) cover
   (b) faceplate
   (c) canopy
   (d) any of these

4. When FMC is used where flexibility is required after installation, _____ shall be installed.

   (a) an equipment grounding conductor
   (b) an expansion fitting
   (c) a flexible nonmetallic connector
   (d) none of these

5. When LFMC is used to connect equipment requiring flexibility after installation, a(n) _____ conductor shall be installed.

   (a) main bonding
   (b) grounded
   (c) equipment grounding
   (d) none of these

6. Snap switches, including dimmer and similar control switches, shall be connected to an equipment grounding conductor and shall provide a means to connect metal faceplates to the equipment grounding conductor, whether or not a metal faceplate is installed.

   (a) True
   (b) False

7. A snap switch that does not have means for connection to an equipment grounding conductor shall be permitted for replacement purposes only where the wiring method does not include an equipment grounding conductor and the switch is _____.

   (a) provided with a faceplate of nonconducting, noncombustible material
   (b) GFCI-protected
   (c) a or b
   (d) none of these

8. Receptacles incorporating an isolated grounding conductor connection intended for the reduction of electrical noise shall be identified by _____ on the face of the receptacle.

   (a) an orange triangle
   (b) a green triangle
   (c) the color orange
   (d) the engraved word "ISOLATED"

9. Isolated ground receptacles installed in nonmetallic boxes shall be covered with a nonmetallic faceplate, unless the box contains a feature or accessory that permits the effective grounding of the faceplate.

   (a) True
   (b) False

10. Receptacles and cord connectors having equipment grounding conductor contacts shall have those contacts connected to a(n) _____ conductor.

   (a) grounded
   (b) ungrounded
   (c) equipment grounding
   (d) neutral

11. Where a grounding means exists in the receptacle enclosure a(n) _____-type receptacle shall be used.

   (a) isolated ground
   (b) grounding
   (c) GFCI
   (d) dedicated

12. Metal faceplates for receptacles shall be grounded.

   (a) True
   (b) False

13. When equipment grounding conductors are installed in panelboards, a _____ shall be secured inside the cabinet.

   (a) grounded conductor
   (b) terminal lug
   (c) terminal bar
   (d) none of these

14. Metal raceways shall be bonded to the metal pole with a(n) _____.

   (a) grounding electrode
   (b) grounded conductor
   (c) equipment grounding conductor
   (d) any of these

15. Exposed conductive parts of luminaires shall be _____.

   (a) connected to an equipment grounding conductor
   (b) painted
   (c) removed
   (d) a and b

16. Where an equipment grounding conductor isn't present in the outlet box for a luminaire, the luminaire must be made of insulating material and must not have any exposed conductive parts.

   (a) True
   (b) False

17. Branch circuits serving patient care areas shall be installed in a metal raceway or cable listed in 250.118 as an effective ground-fault current path.

   (a) True
   (b) False

18. In patient care areas, the grounding terminals of all receptacles and all noncurrent-carrying conductive surfaces of fixed electric equipment _____ shall be connected to an insulated copper equipment grounding conductor.

   (a) operating at over 100V
   (b) likely to become energized
   (c) subject to personal contact
   (d) all of these

19. Metal faceplates for switches and receptacles can be connected to the equipment grounding conductor by means of a metal mounting screw(s) securing the faceplate to a grounded outlet box or grounded wiring device in patient care areas.

   (a) True
   (b) False

20. An equipotential plane is an area where wire mesh or other conductive elements are embedded in or placed under concrete bonded to _____.

   (a) all metal structures
   (b) fixed nonelectrical equipment that may become energized
   (c) the electrical grounding system
   (d) all of these

21. The purpose of the equipotential plane is to prevent a difference in voltage from developing within the plane area.

   (a) True
   (b) False

22. An equipotential plane shall be installed in all concrete floor confinement areas of livestock buildings, and all outdoor confinement areas with a concrete slab that contain metallic equipment accessible to livestock and that may become energized.

    (a) True
    (b) False

23. Equipotential planes shall be installed in outdoor concrete slabs where metallic equipment is located that may be energized and is accessible to livestock, other than poultry.

    (a) True
    (b) False

24. The equipotential plane in an agricultural building shall be connected to the electrical grounding system with a copper, insulated, covered, or bare conductor and not smaller than _____.

    (a) 10 AWG
    (b) 8 AWG
    (c) 6 AWG
    (d) 4 AWG

25. Metal equipment racks and enclosures for permanent audio system installations shall be grounded.

    (a) True
    (b) False

26. Exposed noncurrent-carrying metal parts of an information technology system shall be _____.

    (a) bonded to an equipment grounding conductor
    (b) double insulated
    (c) GFCI-protected
    (d) a or b

27. An 8 AWG or larger solid copper equipotential bonding conductor shall be extended to service equipment to eliminate voltage gradients in the pool area.

    (a) True
    (b) False

28. The _____ pool bonding conductor shall be connected to the equipotential bonding grid either by exothermic welding or by pressure connectors in accordance with 250.8.

    (a) 8 AWG
    (b) insulated or bare
    (c) copper
    (d) all of these

29. The pool structure, including the reinforcing metal of the pool shell and deck, shall be bonded together.

    (a) True
    (b) False

30. Which of the following shall be bonded?

    (a) Metal parts of electrical equipment associated with the pool water circulating system.
    (b) Pool structural metal.
    (c) Metal fittings within or attached to the pool.
    (d) all of these

31. Metal conduit and metal piping within _____ horizontally of the inside walls of the pool shall be bonded.

    (a) 4 ft
    (b) 5 ft
    (c) 8 ft
    (d) 10 ft

32. Metal parts of electric equipment associated with an indoor spa or hot tub water circulating system shall be bonded.

    (a) True
    (b) False

33. Metal raceways and metal piping within _____ of the inside walls of an indoor spa or hot tub, and not separated from the indoor spa or hot tub by a permanent barrier, shall be bonded.

    (a) 4 ft
    (b) 5 ft
    (c) 7 ft
    (d) 12 ft

34. Metal parts associated with an indoor spa or hot tub shall be bonded by _____.

    (a) the interconnection of threaded metal piping and fittings
    (b) metal-to-metal mounting on a common frame or base
    (c) a solid copper bonding jumper not smaller than 8 AWG solid
    (d) any of these

35. Metal piping systems and all grounded metal parts in contact with the circulating water of a hydromassage bathtub shall be bonded together using a(n) _____ solid copper bonding jumper not smaller than 8 AWG.

    (a) insulated
    (b) covered
    (c) bare
    (d) any of these

36. The 8 AWG solid bonding jumper required for hydromassage bathtubs shall not be required to be extended to any _____.

    (a) remote panelboard
    (b) service equipment
    (c) electrode
    (d) all of these

37. The metallic sheath members of communications cable entering or attached to buildings shall be _____.

    (a) grounded at the point of emergence through an exterior wall
    (b) grounded at the point of emergence through a concrete floor slab
    (c) interrupted as close to the point of entrance as practicable by an insulating joint
    (d) any of these

38. In one- and two-family dwellings, the primary protector grounding conductor for communications systems shall be as short as practicable, not to exceed _____ in length.

    (a) 5 ft
    (b) 8 ft
    (c) 10 ft
    (d) 20 ft

39. In one- and two-family dwellings where it's not practicable to achieve an overall maximum primary protector grounding conductor length of 20 ft, a separate ground rod not less than _____ shall be driven and it shall be connected to the power grounding electrode system with a 6 AWG conductor.

    (a) 5 ft
    (b) 8 ft
    (c) 10 ft
    (d) 20 ft

40. Limiting the length of the primary protector grounding conductors for communications circuits helps to reduce voltage between the building's _____ and communications systems during lightning events.

    (a) power
    (b) fire alarm
    (c) lighting
    (d) lightning protection

41. For buildings with grounding means but without an intersystem bonding termination, the grounding conductor for communications circuits shall terminate to the nearest _____.

    (a) building or structure grounding electrode system
    (b) interior metal water piping system, within 5 ft from its point of entrance
    (c) service equipment enclosure
    (d) any of these

42. Communications electrodes must be bonded to the power grounding electrode system using a minimum _____ copper bonding jumper.

    (a) 10 AWG
    (b) 8 AWG
    (c) 6 AWG
    (d) 4 AWG

43. Each lead-in conductor from an outdoor antenna shall be provided with a listed antenna discharge unit, unless enclosed in a grounded metallic shield.

    (a) True
    (b) False

44. Antenna discharge units shall be located outside the building only.

    (a) True
    (b) False

45. The grounding conductor for an antenna mast shall be _____ protected where subject to physical damage.

    (a) electrically
    (b) mechanically
    (c) arc fault
    (d) none of these

46. The grounding conductor for an antenna mast or antenna discharge unit shall be run to the grounding electrode in as straight a line as practicable.

    (a) True
    (b) False

47. If the building or structure served has an intersystem bonding termination, the grounding conductor for an antenna mast shall be connected to the intersystem bonding termination.

    (a) True
    (b) False

48. The grounding conductor for an antenna mast or antenna discharge unit, if copper, shall not be smaller than 10 AWG.

    (a) True
    (b) False

49. If a separate grounding electrode is installed for the radio and television equipment, it shall be bonded to the building's electrical power grounding electrode system with a bonding jumper not smaller than _____ AWG.

    (a) 10
    (b) 8
    (c) 6
    (d) 1/0

50. The outer conductive shield of a CATV coaxial cable entering a building shall be grounded as close to the point of entrance as practicable.

    (a) True
    (b) False

51. The conductor used to ground the outer cover of a CATV coaxial cable shall be _____.

    (a) insulated
    (b) 14 AWG minimum
    (c) bare
    (d) a and b

52. In one- and two-family dwellings, the grounding conductor for CATV shall be as short as practicable, not to exceed _____ in length.

    (a) 5 ft
    (b) 8 ft
    (c) 10 ft
    (d) 20 ft

53. In one- and two-family dwellings where it's not practicable to achieve an overall maximum grounding conductor length of _____ for CATV, a separate grounding electrode as specified in 250.52(A)(5), (6), or (7) shall be used.

    (a) 5 ft
    (b) 8 ft
    (c) 10 ft
    (d) 20 ft

54. Limiting the length of the primary protector grounding conductors for community antenna television and radio systems reduces voltages between the building's _____ and communications systems during lightning events.

    (a) power
    (b) fire alarm
    (c) lighting
    (d) lightning protection

55. The grounding conductor for CATV coaxial cable shall be _____ protected where subject to physical damage.

    (a) electrically
    (b) arc fault
    (c) physically
    (d) none of these

56. A bonding jumper not smaller than _____ copper or equivalent shall be connected between the CATV system's grounding electrode and the power grounding electrode system at the building or structure served where separate electrodes are used.

    (a) 12 AWG
    (b) 8 AWG
    (c) 6 AWG
    (d) 4 AWG

Index